Getting Rid of Graffiti

OTHER TITLES FROM E & FN SPON

Amenity Landscape Management
Edited by R. Cobham

Building Failures: Diagnosis and Avoidance
W.H. Ransom

City Centre Planning and Public Transport
B.J. Simpson

Countryside Management
P. Bromley

Defects and Deterioration in Buildings
B.A. Richardson

Durability of Building Materials and Components
Edited by J.M. Baker, P.J. Nixon, A.J. Majumdar and H. Davies

Environmental Planning for Site Development
A.R. Beer

Grounds Maintenance
P. Sayers

Hazardous Building Materials
S.R. Curwell and C.G. March

A Handbook of Segmental Paving
A. Lilley

Housing Defects Reference Manual
Building Research Establishment

The Maintenance of Brick and Stone Masonry Structures
Edited by A.M. Sowden

Parking: A Handbook of Environmental Design
J. McCluskey

Planning and the Heritage
M. Ross

Protection of Concrete
Edited by R.K. Dhir and J.W. Green

Safety in the Built Environment
J.D. Sime

The Soiling and Cleaning of Building Facades
Edited by L.G.W. Verhoef

Spon's Grounds Maintenance Contract Handbook
R.M. Chadwick

Spon's Landscape Handbook
Derek Lovejoy and Partners

For more information on these and other titles please contact:
The Promotions Department, E & FN Spon, 2–6 Boundary Row, London SE1 8HN
Telephone 071-865 0066

Getting Rid of Graffiti

A practical guide to graffiti removal and anti-graffiti protection

M.J. Whitford

BTech(Hons), MSc, PhD, CChem, MRSC
Materials Chemistry Section,
Scientific Services Division,
London Underground Ltd

With a Foreword by Professor G.W. Ashworth,
Director General, The Tidy Britain Group

E & FN SPON
An Imprint of Chapman & Hall
London · New York · Tokyo · Melbourne · Madras

**Published by E & FN Spon, an imprint of Chapman & Hall,
2–6 Boundary Row, London SE1 8HN**

Chapman & Hall, 2–6 Boundary Row, London SE1 8HN, UK

Van Nostrand Reinhold Inc., 115 5th Avenue, New York NY10003, USA

Chapman & Hall Japan, Thomson Publishing Japan,
Hirakawacho Nemoto Building, 7F, 1-7-11 Hirakawa-cho,
Chiyoda-ku, Tokyo 102, Japan

Chapman & Hall Australia, Thomas Nelson Australia, 102 Dodds Street,
South Melbourne, Victoria 3205, Australia

Chapman & Hall India, R. Seshadri, 32 Second Main Road, CIT East,
Madras 600 035, India

First edition 1992

© 1992 London Underground Limited

Typeset in 11/13pt Sabon by EJS Chemical Composition, Bath
Printed and Bound in Great Britain by Hartnolls Limited, Bodmin, Cornwall.

ISBN 0 419 17040 5   0 442 31490 6   (USA)

A catalogue record for this book is available from the British Library.

Library of Congress Cataloging-in-Publication Data available.

Warning Whilst every effort has been made to describe effective ways of removing graffiti from surfaces, materials, etc., neither London Underground Ltd, nor the author accept responsibility nor can they be held liable for any loss or damage suffered by any person or to any property in the event that the user fails to strictly follow the instructions set out within this work, observes all proper necessary safety conditions and complies with all health and safety legislation, in particular (in the United Kingdom), *The Control of Substances Hazardous to Health (COSHH) Regulations* (see Appendix E) issued from time to time. In case of any doubt, the user should seek special advice before implementing any procedures described herein.

Contents

Colour plates appear between pages 80 and 81

Contents

Contents

Foreword

Whether or not graffiti is art and whether or not it is therapeutic for the perpetrator may, for some, be a fascinating topic for debate. Let those who wish to exercise their intellect on such matters do so. For people faced with graffiti day after day as they leave their homes, travel to work, escort children to school, take the subway to the shops, travel by train, pass a public monument or use a public loo, an academic approach is irrelevant. Even as a medium for the expression of political disillusionment or defiance graffiti has to be low on the list of effective options. Today's incisive political sideswipe quickly becomes anachronistic and boring.

The police regard graffiti spraying as the second criminal act in a line that leads from littering through graffiti to vandalism and violence. The overwhelming majority of the rest of us see graffiti as offensive, and when it appears want it removed completely and permanently from the scene. All too often the hope that it will not reappear is in vain, but that is no reason not to tackle the onerous task of removal. And onerous it is.

My time as Director General of the Tidy Britain Group has brought me into contact with voluntary organizations of all kinds. Many environmental improvement tasks can be carried out effectively, and enjoyably, by volunteers. I do not recommend graffiti removal as belonging to that category. It is a job for professionals. It usually involves the use of hazardous chemicals and, in some locations, carries a risk to life and limb. It calls for technical know-how; a badly carried out job can leave a surface looking as unsightly as before. In many cases graffiti removal falls upon the local authority and as councils increasingly put graffiti removal out to competitive tender, specialist companies are emerging to take on the role.

In producing this book, therefore, Dr Whitford is doing a service to all of us. For whether we suffer enforced exposure to graffiti in our daily lives, whether we have a campaigning role to improve local environments or whether we carry a responsibility for graffiti removal, the anti-graffitists can now fight back! Dr Whitford provides a uniquely comprehensive source of technical information, methodology, safe working practices, products and equipment. Having an architectural background, I am well aware of the importance of appropriate materials specification and I applaud Dr Whitford's attention to the graffiti resistance of surfaces, long-term protection of surfaces and aesthetic considerations.

Urban living has much to commend it and I recognize that a sterile environment will not satisfy the need within us for the vibrancy and character that modern towns and cities provide. Though graffiti does indicate a human presence I believe that, along with litter and fly-posting, it is endured rather than accepted, invariably creating an atmosphere of neglect and unease rather than 'life'. Where it is threatening, it can be a portent of worse to come.

I hope that those responsible for public areas and facilities will eagerly put to good use the wisdom and expertise contained within this book. For whilst we cannot always be one-step ahead of the graffitists, we can now at least ensure we are hot on their heels.

Professor Graham W. Ashworth CBE
Director General of The Tidy Britain Group

Preface

This book owes its existence to the work carried out over a period of years by London Underground Limited on methods of removing, and other means of counteracting, graffiti. The Underground, like the New York subway and similar systems in major cities around the world, is a prime target for producers of graffiti (whom we call 'graffitists' throughout this volume).

Since its property includes trains, buildings, platforms, passageways, concourses, waiting rooms, public lavatories, bridges, other structures and more than 250 miles of rail track, the Underground has as great a variety of vulnerable surfaces as any other authority in Great Britain, and the amount of graffiti to which it is subjected is among the largest of any enterprise in the world.

Initiatives by London Underground to combat the problem have included reviews of security at train depots, the installation of surveillance equipment, the use of security patrols, a study of the psychology of graffitists and the impact of graffiti on passengers, liaison with other sections of London Regional Transport and consultation and exchange of information with other authorities both in the United Kingdom and abroad. However, the most intensive and sustained anti-graffiti activity has consisted of:

- laboratory research into the nature of graffiti markings and the means of getting rid of them;
- site trials on graffiti removal and the protection of vulnerable surfaces;
- the development of substances, methods, equipment and instructions for the removal of graffiti from various surfaces;
- research into protective measures and graffiti-resistant surface materials and coatings;
- training and deployment of staff to remove graffiti.

The bulk of this book is devoted to describing in some detail the methods and procedures that have been found to achieve the best results in graffiti removal. Although these methods were developed for London Underground, the techniques can be used elsewhere. Better removal agents and equipment may come on to the market in future, and different formulations for graffiti-removal agents may be needed when graffitists get hold of new types of marking material, but we are confident that the methods set out in the chapters that follow are, in general, at least as effective as any in use elsewhere. What is important is to ensure that the operators engaged in the process of graffiti removal use only approved materials, equipment and methods. It is not uncommon for operators to ignore instructions, use unauthorized types of solvent, neglect quality standards, fail to clean up after completing the work, depart from laid down procedures and find a host of ingenious ways to circumvent safety rules. Departures from approved procedures result in inadequate cleaning, often leaving the surface looking worse than before work began on it. In addition, by neglecting to obey safety instructions, operators put themselves and others at risk. Sound training and good supervision are essential for a successful anti-graffiti operation.

Acknowledgements

This book is based on a manual, written for internal use by London Underground Limited, entitled *Reference Manual for Graffiti Removal and Surface Protection*. As Section Head of the Materials Chemistry Section, I was responsible for the production of the manual and for the development of chemicals, equipment, methods and instructions that are now used by London Underground to get rid of graffiti.

However, the manual would have been of less value and the laboratory work less relevant if this study had been confined to the laboratory. By going out to graffiti sites, frequently late at night, finding out for myself who the people are who produce the markings, what motivates them and how they are organized, by examining their equipment and materials, and by talking to many other people concerned with the problem of graffiti, I have tried to develop an understanding of the subject which I hope may give this work practical value.

With regard to the present book, the knowledge so gleaned would not have been as easy for the user to understand were it not for Henry Galgut of 'Plain Words', whose excellent copy-writing skills and intuitive grasp of the subject have transformed the information into a readable reference work, which I believe will stimulate all those with an interest in the subject. Also what is essentially a 'visual' topic would be barren without the excellent photos provided by the Photographic Service Section of London Underground and the less artistic but none the less useful 'snaps' provided by my trusty 'Nikon'.

I am also grateful to Rod Brans, Scientific Adviser (Applied Chemistry) who has overall responsibility for the Materials Chemistry Section amongst others, for ensuring the acceptability of the contents of the book, and to Nick Lewis and Jeff Levy, my regular contacts in the London Underground Advertising and Publicity Department, who organized the publication of the work.

None of this would have been possible without the sheer hard work of members of my own Section and the unstinting collaboration of other London Underground Divisions, Transport Authorities both in Britain and overseas, Council Authorities, specialist cleaning restoration and graffiti removal companies, product and equipment suppliers and the general public. Names are too numerous to mention and if I tried I would certainly commit the cardinal sin of omitting some. They know who they are and hopefully will take justifiable pride in the part they are playing in *Getting Rid of Graffiti*.

Maurice Whitford Ph.D.

1 History and trends in graffiti

1.1 Definitions and vocabulary

Graffiti is a plural word for drawings, patterns, scribbles or messages painted, written or carved on walls and other surfaces. The singular form of the word is 'graffito' from the Italian 'graffio' – a scratch – but most graffiti take the form of surface scribbles and designs. In this book, we use the word 'graffitist' to describe the people who produce graffiti.

Graffitists' slang

Bite	To copy another's graffiti style
Bomb	To apply graffiti intensively to a location
Buff	To erase. Any means adopted by the 'authorities' to remove graffiti
Burner	A great piece
Caps (fat or skinny)	Interchangeable spray can nozzles allowing variation of spray width
Crew	Loosely organized group of graffitists (also 'clique')
Dog	To overwrite another's graffiti
Get up	Become known among graffiti crews
Hit	To tag or bomb a surface
King/Queen	Dominant graffitist on bus route or underground railway line
Piece	Coloured, complex pictorial graffito in spray paint (from 'masterpiece')
Tag	Graffitist's stylized 'signature' in marker or spray paint
Throw-up	Quickly executed graffiti in spray paint

| Toy | Poor work; also, inexperienced or incompetent graffitist |
| Writer | Graffitist |

1.2 The evolution of graffiti

Graffiti have been around as long as human society. The paintings and engravings of animals in the Lascaux caves in the Dordogne region of France are said to date from 18 000 BC. Similar images, painted thousands of years ago on cave walls, rock faces and boulders, still exist in Asia, Africa and Australia. Graffiti, many of them obscene or political, are found on ancient Egyptian monuments, on the walls of the city of Pompeii and in many other parts of the world. At all times down the centuries, innumerable people have given way to the temptation to leave their marks on walls, trees, school desks and any blank surfaces that attract them. Now, obscenities and witticisms on public lavatory walls and scrawls in other public places are commonplace. They are usually the work of individuals and are tolerated, if not approved of, by most people.

However, in the last few years, there have been dramatic changes in the extent and type of graffiti and the motivation of those responsible for them. Most modern graffiti originated in depressed areas of New York as 'street art' and spread to the capital cities of Europe, particularly London, Paris, West Berlin and Amsterdam, and other cities around the world. The followers of the new international graffiti cult, with its own language, profess to be practising a form of genuine art.

The New York graffitists made the exteriors of subway trains their first targets. So many people were engaged in the intensive and prolonged assault on the trains that the sheer volume of graffiti frustrated all efforts to clean it off and defied all attempts at deterrence through law enforcement. In London and other cities, some local authorities conferred a degree of respectability on certain talented graffitists by commissioning them to paint murals in public places. One New York graffitist, Jean-Paul Basquiet, collaborated with Andy Warhol throughout 1984 and 1985 to produce joint paintings which filled three London galleries at the same time in December 1988 and January 1989.

Ventures that encourage graffitists may get the approval of

people who believe that anything produced with a hand-held instrument has a claim to be considered art, but to those whose property is subjected to the unsolicited attentions of graffitists – no matter how talented – the markings are a form of vandalism, often distressing to the viewers and expensive to remove.

Graffitists are encouraged by anything that appears to condone their activities. Evening classes and lessons in youth clubs in graffiti techniques, the use of graffiti styles in advertising, media coverage of 'street art' and invitations to join the art establishment all tend to stimulate the proliferation of graffiti in locations where they are not welcome. Leniency on the part of those magistrates who do not equate the execution of graffiti with vandalism also does nothing to discourage the practice.

With the widespread popularization of 'graffiti art', there has been an alarming escalation of graffiti marking in public places and on trains and buses. The invention of aerosol spray paints enabled a few people to produce reasonably artistic 'pieces', but also encouraged the proliferation of much less accomplished work throughout the Western world. In addition, the development of chemical products that have contributed to improvements in the quality of paints has enabled graffitists to produce longer-lasting work that challenges the ingenuity of those who have to remove it.

A large number of gangs ('crews'), mostly of young people, has emerged. They compete to mark out their territory and establish domination over rival crews by the use of distinctive symbols ('tags'). Most major cities, and many smaller towns, are at the receiving end of an increasing barrage of unsightly tags that by no stretch of the imagination can be called art, the main aim in their creation being speed and frequency of execution. Felt tip markers and aerosol spray cans are most commonly used to produce the markings.

Graffitists have an insatiable appetite for attacking trains and buses which provide a mobile, highly-visible platform for their tags. In London, both trains and buses on certain routes and some garages, depots and a wide variety of other buildings and structures are subjected to repeated attacks, often by the same crews, judging by the tags. Other cities suffer in the same way. It is demoralizing both for the public, who think nothing is being done about the problem, and for cleaning staff who feel their efforts are in vain. Crossing high-voltage rails and even positioning themselves in alcoves in underground tunnels to spray moving carriages, gives some graffitists a sensation of danger which adds to the satisfaction they get from leaving their marks on rolling stock. A number have been killed painting or spraying in underground systems.

Some of the work of graffitists equipped with spray guns has also been extremely dangerous to the public. Signs and warning lights in transport systems have been sprayed over to obliterate them. Occasionally too, automatic ticket machines have been sprayed, creating a nuisance to travellers.

Originally, the outside surfaces of buses and trains were the prime targets for graffiti, but soon the practice spread to the interiors, and there was a considerable increase in graffiti attacks on public transport premises as well. The transport services as a whole were singled out as the graffitists' foremost enemy.

1.3 The psychology of graffitists

The Lascaux cave paintings are thought to have had a magical significance concerned with a hunting ritual. Perhaps something of that psychological effect animates the subconscious of some graffitists. Certainly, many of them roam around in packs ('crews') and the markings they make are invested with a significance meaningful both to themselves and rival crews.

Graffiti styles vary according to the motivation of their creators. Some types are categorized as graffiti art, some are political, some have a sporting bias, others are racial, personal, sexual, obscene or simply mindless. Many tags are simply highly stylized nicknames. In Sydney, Melbourne, London and other cities, crews of women have gone around writing slogans, many of them undeniably witty, on posters that they perceive to be either exploiting women or promoting a 'macho' image of men.

Graffiti crews are highly organized and hierarchical in structure. An individual graffitist may wish to be invited to join a crew. To succeed in that ambition ('to get up'), he or she has to produce tags or pieces that are accepted as stylish, well-executed and with a recognisably personal identity. If the work is carried out in difficult or dangerous locations, so much the better.

Typically, crews consist of between four and ten people, often of mixed races and aged between 14 and 20 who work together, led by a 'king' or 'queen'. Sometimes crews contain a member trained in graphic design who helps them develop new designs. Sometimes they photograph their work and circulate the results among other crews to support their claims to be the most accomplished. Usually, a crew will acknowledge good work and accept that the only way to gain recognition as the superior crew is to produce better pieces or tags. It used to be generally considered wrong to deface ('dog') another crew's work, but the practice is now more common, with results that are even more unsightly than the original markings. Inferior graffiti is sometimes denounced with the word 'toy' or 'try' (Plate 1). A crew may break up if the leader leaves it to take up a respectable job or because he or she has been arrested.

There are thought to be at least 300 graffiti crews operating in London alone. Most do not regard themselves as vandals, although some are motivated by vindictive feelings against the authorities. Many believe they are improving the environment by enlivening drab or derelict surroundings with their bright, coloured designs. They understand that many people find graffiti offensive and anti-social, but they do not accept that their actions are serious crimes. On the other hand, there are some graffitists whose motives are deliberately anti-social, who set out to alarm and anger through ugly, violent or obscene imagery or wording.

Interviews with graffitists have revealed that they get satisfaction from indulging in an unlawful pursuit and in beating the system. Some are stimulated by the spice of danger that comes from leaving their marks in unsafe places. Many are hungry for the recognition they believe they can get in no other way. Some are just saying 'I'm here' to the world or to others of like mind who may feel inadequate or neglected.

1.4 Public perception of graffiti

Research into the effects of graffiti on the public has revealed that although some people are indifferent to them, most regard them as a form of vandalism contributing to the degeneration of the environment (Plate 2). Scribbling and tags are generally disliked, particularly inside buses and trains where they cannot be ignored (Plate 3).

Nearly everyone takes exception to obscene, racist and sexist graffiti and some people even find them menacing. Most graffitists are either unconcerned about the effects of their handiwork or deliberately set out to disturb those who have to look at it. They aim for maximum impact through the use of high-contrast sprays and pens, large tags and pieces in highly visible locations (Plate 4), widespread replication of individual tags and, in some cases, offensive slogans and statements. The result is that many people feel as though they are living in an uncontrolled and frightening environment.

Intensive applications of graffiti, whether well- or badly-executed, terrify some people. In the New York subway system, the number of people using the subway fell as the volume of graffiti increased. Between 1981 and 1983, when the problem was at its height and nearly every train and subway station was covered in graffiti, the number of passengers fell by 12%. There was no improvement in usage of the system until 1985 when a comprehensive programme of cleaning, policing and community education was introduced. Then, as the programme was seen to be succeeding, passengers began to return to the subway. Now that the system is almost free of graffiti, more people than ever travel on it. Usage has increased by over 30%.

The majority of people make a distinction between carefully-executed colourful pieces on walls and untidy, indiscriminate graffiti. A colourful, inoffensive and carefully-executed mural ('piece') on an otherwise dull wall is more tolerable than crude tagging of public buildings. In Paris, an outbreak of small stencilled designs, repeated patterns of animals, birds, dancers and a variety of other shapes and images, were regarded by many people as harmless and even rather charming. But even a colourful piece in the wrong place is unwelcome, and the work of manifestly untalented or malicious scribblers is almost universally detested. Unfortunately, it is that type of graffiti which predominates.

The effect of graffiti on the public at a particular location may not be proportionate to its intensity. While an intense attack on a heavily-used train station would probably have an appreciable impact on the travellers there, an attack of similar intensity on a secluded wall would obviously make little impression on people. On the other hand, even a small mark that would normally pass unnoticed, could be judged intolerable on a new surface or a distinguished building.

Apart from their obvious adverse impact on the environment, graffiti sometimes have an undesirable indirect effect. Some authorities, unable to overcome the problem, and unwilling to waste money on expensive materials that are likely to become defaced by graffiti, have specified cheaper and shoddier construction materials for new projects. So what might have been attractive new buildings and precincts are deliberately designed to be tawdry and unappealing.

Graffitists are sometimes caught and punished. Increased vigilance and patrolling of vulnerable areas may lead to more arrests and deter some would-be graffitists. Identification of tags and graffiti style may help in tracking down culprits. Speedy and thorough removal of graffiti may discourage some people. However, the problem is not likely to disappear quickly. Perhaps the development of new removal agents and protective materials, more consultation between affected parties, better surveillance, quicker response times and improved methods of graffiti removal will, in time, ameliorate the problem.

2 Anti-graffiti policy

It is advisable for any organization whose property is likely to be hit by graffiti to have a written policy on graffiti removal, surface protection and other countermeasures. The internal publication of such a policy clarifies the organization's stance on the problem and lets employees know how they are expected to deal with graffiti, and what they should do if they come across anyone defacing property.

The nature and circumstances of individual organizations will determine much of the content of their anti-graffiti policies, but most will contain statements along the following lines.

- Graffiti must be removed as soon as possible after it has been discovered, bearing in mind the overriding necessity to keep the business running.
- Health and Safety Regulations must be followed in all graffiti removal and graffiti protection work.
- Only materials and methods approved by the company may be used in anti-graffiti operations.
- Efforts should be made to remove highly visible and more disagreeable graffiti before less offensive matter.
- In general, the order of priority for removal should be
     obscene and racist graffiti;
     personal graffiti, such as names and telephone numbers;
     libellous or defamatory graffiti;
     other graffiti;
  but where graffiti has a high visual impact, its removal should be given a high priority regardless of its content.
- Where graffiti removal or surface protection work could interfere with, or create a hazard for, the public or staff, the work area must be securely fenced off.
- Hazardous operations must be continuously supervised and all work must have adequate control.

- The protection of the public is the prime consideration at all times.
- Members of staff discovering graffitists at work should immediately contact their supervisor who will inform the appropriate security personnel.

3 Assessing the problem

It is necessary, from time to time, and especially when setting up a working system for graffiti removal, to carry out inspections of affected areas, and when surveying a site, it is useful to make a record of the location and nature of the graffiti.

The record could include the following points.

- The location of the graffiti.
- Any other areas that might attract graffiti.
- Type and extent of the graffiti.
- Surface types affected.
- Surface types at risk from graffiti attack.
- Public access to the affected area.
- Feasibility of isolation of the area for graffiti removal or surface protection.

Colour photographs may be taken and attached to the record; these give the clearest indication of the extent of the problem. It may even be helpful to mark drawings or sketches of the site to show the areas affected.

Of course, it is not necessary to follow the whole of this procedure every time a graffiti removal operation is undertaken, but a record does define the problem. It provides a basis for deciding what procedures, materials and methods to use in attacking the problem and what steps to take to forestall or counteract any further appearance of graffiti.

Solutions to the problem may demand the use of chemical substances, protective clothing and advanced methods. On the other hand, the counteractive methods may be relatively simple. For example, if poster advertisements are defaced, it may be possible to overstick promptly with new posters. More permanent posters could be made of graffiti-resistant materials or be protected inside display cases constructed from materials resistant to graffiti.

Similarly, for graffitied surfaces where effective cleaning methods are already established, e.g. on ceramic tiles, the solution may be to increase the frequency of cleaning or to find even better removal agents and methods – improved products are constantly coming onto the market. There is evidence to suggest that graffitists turn their attentions elsewhere if their tags are quickly and completely removed.

After completion of graffiti removal or protection work, records should be annotated to show the results and any lessons learned.

3.1 Trial procedures

In difficult cases, it may be necessary to visit the site to:

- agree where to carry out graffiti removal, and possibly surface-protection, trials;
- supervise such trials;
- assess completed trials;
- inspect trial areas to evaluate effectiveness;
- assess intended large scale removal or protection programmes.

Whether or not trials are deemed necessary, good supervision and monitoring of the work are always necessary.

3.2 General considerations

Having obtained information from site visits and the written and photographic records, a number of other factors may need to be taken into account before tackling graffiti. These may be:

- the organization's graffiti removal or protection policy;
- aesthetic considerations;
- statutory considerations (e.g. listed buildings);
- costs;
- when to do the work (e.g. when the public is not present);
- deployment of staff and equipment;
- methods and materials;
- alternative methods of combating the problem (e.g. policing, surveillance, security systems);
- Health and Safety regulations and Codes of Practice.

3.3 Location of target surfaces

It is important to be precise about the location of surfaces for graffiti removal or protection to enable decisions to be made as to which techniques and materials are the most appropriate to use.

The choice of techniques and materials may be influenced by the availability of services such as water and electric power. Permission may have to be obtained for isolation of the whole or part of the area affected. The effect on the public or employees of partitioning areas off during both graffiti removal and the application of protective coatings will also have to be taken into consideration.

Health and Safety monitoring may be required during the work to ensure neither staff nor the public are at risk from fumes, fire, noise or other hazards.

4 Graffiti marker types

In principle, any device or material that will mark a given surface in a controllable way, may be used by a graffitist. Table 4.1 shows some of the marking devices most commonly used.

Because graffitists want to produce their marks quickly, boldly and indelibly, the devices they prefer to use nowadays are aerosol spray paints of various compositions, and waterproof ink broad-nib felt-tip markers. While ball-point pen, lipstick, brush-applied paints and scratching are all used from time to time, their use is insignificant compared with spray paints and felt-tip markers.

When assessing a specific graffiti problem, it is important to identify the general marker type, or types, correctly. The method of removal will depend on the chemical composition and susceptibility to removal agents of the markers, and the extent to which they have penetrated the surface. In addition, the marker type and graffiti style may give warning of new trends in graffiti attack and, together with the tag form, may help in the tracking down of the individual or crew responsible.

Although Table 4.1 suggests that specific marker types are favoured for particular surfaces, a mixture of marker types is often used, particularly in places exceptionally prone to graffiti attacks, such as the 'Halls of Fame' that exist in some urban locations. Mixtures also crop up where a competing crew has overwritten another crew's efforts ('dogged their tags'). In such cases, all marker types must be identified and expert advice obtained to get the best sequence and techniques of removal. Colour photographs are useful for such assessments.

4.1 Felt-tip marker pens

Most felt-tip marker pens used by graffitists are of the solvent-based 'permanent' type. The marks made by these pens resist water and

Graffiti marker types

Table 4.1 Graffiti marker types and target surfaces

Marker/Device type	Typical target surface	Removability
1 Aerosol paints Acrylic, cellulose and other bases; pigmented, metallic, fluorescent and other finishes	All surface types (including very rough); very widely used for 'street art'	Varies according to surface attacked. Very difficult for rough surfaces
2 Paints (brush-applied) Including domestic and rubber-based underseal	All surfaces; limited use mainly for 'political' or 'sport' graffiti in exterior locations	Varies according to surface attacked. Very difficult for rough surfaces
3 Felt-tip pens (i) 'Permanent' (solvent-based)	Mainly smooth/semi-smooth; very widely used for 'street art'	Fairly easy for non-permeable/non-porous surfaces, moderate/difficult for permeable/porous surfaces
(ii) Non-permanent (water-based)	Smooth/semi-smooth; little used	Easy from most surfaces
4 Ball-point pen	Smooth surface, mainly small-scale graffiti by individuals	Easy from non-permeable/non-porous surfaces; moderate/difficult for permeable/porous
5 Lipstick/Wax crayon	Usually smooth surfaces, little used	Moderate/difficult for porous/permeable surfaces
6 Chalk	Various surfaces, little used	Easy
7 Pencil	Smooth surfaces, little used	Usually easy
8 Knife/Other scratching implement	All surfaces/textures, mainly soft or crumbly materials; mainly small-scale graffiti by individual	Difficult – surface may require filling or re-finishing
9 Self-adhesive stickers	Smooth surfaces; variable incidence, often 'political' or 'social' comment	Easy (hard surfaces) moderate/difficult (soft, e.g. plastic surfaces)

Table 4.1 *Continued*

Marker/Device type	Typical target surface	Removability
10 Pasted 'fly-posters'	Smooth/semi-smooth exterior surfaces; widely used for illicit publicity	Usually easy

detergents, unlike marks of water-based felt-tip markers. Determined graffitists favour the most permanent of 'permanent' markers, and use the widest-nibbed ones they can find to achieve high visibility for their tags. Not only do they use such markers for quickly-executed tags ('throw-ups'), but also to outline larger graffiti ('pieces') which are then filled in with aerosol spray paint.

The 'permanence' of the marks may derive from high loading of pigment or dyestuff in the ink, the type of dyestuff or colouring, or the ink's solvent and binder base. Dedicated graffitists have even been known to mix their own inks and make extra wide nibs to achieve their aims.

4.1.1 Felt-tip graffiti removal

Success in removing felt-tip graffiti marks depends largely on the nature of the surfaces on which they have been made. Very smooth non-permeable and non-porous surfaces, such as stainless steel, glazed ceramic tiles, plain glass and some melamines may be easily cleaned with appropriate solvent-based graffiti removal agents.

Liquid inks, however, migrate into the permeable and porous surfaces of materials such as vinyl, conventional paints and terrazzo, making removal more difficult, if not impossible. In such cases, solvent-based removal agents must not be used as they may chase the graffiti dyes further into the surface, leaving unsightly ghosting or smears. A suitable bleaching product may, however, erase felt-tip marks from certain porous surfaces – for example, from terrazzo, bricks and concrete. Marks made by heavily-loaded, metallized and other felt-tip markers may also require mechanical scrubbing to take non-bleachable components out of the ink. And some bleach-resistant dyestuffs may need a further application of the bleaching substance for complete eradication, provided it does not also bleach the surface. Sanding, grit-blasting and other mechanical removal methods should be used only in special circumstances as they may destroy the original surface finish.

4.2 Aerosol spray paints

Aerosol spray paints, lacquers and colourants are extremely popular with graffitists because of their following attributes.

- They can be used on any surface, regardless of texture, since the marking device does not have to be in contact with the surface.
- Surface coverage can be very rapid, minimizing the risk of detection; so large areas may be painted quickly, allowing bold tags and pieces to be executed.
- Control of graffiti mark width and intensity is easily achieved by substituting 'fat' and 'skinny' caps on the same can, or by altering spray distance. Paint loading can be high.
- Drying time is very short, reducing the chances of fresh graffiti removal, particularly on rough or heavily textured surfaces.
- A very wide range of colours, fluorescent shades and metallic finishes are available, making for maximum contrast and high visibility, and easy identification of the individual's graffiti.
- An extensive range of spray paint compositions are on the market, enhancing the chances of some being resistant to a universal removal agent, and maximizing the probability of survival of the graffiti.

In the USA, graffitists are increasingly resorting to certain polyurethane paints which are likely to be particularly resistant to solvent removal. It may be only a matter of time before such American practices are imitated by graffitists in the UK and in other countries. In that event, special removal and protection methods will be required. New protective coatings or removal techniques are constantly under development which may alleviate the problem.

4.2.1 Aerosol paint graffiti removal

Aerosol paint graffiti are usually easily eradicated from smooth non-permeable surfaces, such as glass, glazed ceramic tiles, some melamines and stainless steel, by a solvent-based graffiti removal agent, followed by washing down. Even some permeable surfaces, such as smooth terrazzo, may be effectively cleaned in the same way, provided that the solvent cleaner is not allowed to soak into the surface after softening the paint which, unlike felt-tip marks, largely dries on the surface. A trial should first be done on a small unobtrusive area to test the efficacy of the method.

Permeable surfaces, such as paints and some soft plastic claddings, that are sensitive to the solvents found in graffiti removal products should not be cleaned with such agents without expert advice.

Aerosol graffiti can be largely removed from unpainted and weathered aluminium by a solvent-based graffiti removal gel and medium-pressure water jet washing.

4.3 Brush-applied paint

Paint applied by brushes is usually found on the exteriors of buildings and other, often isolated, structures. The incidence and location of this type of marking may dictate the method and equipment to be used to remove it.

In general, considerations and methods of removal similar to those for aerosol paint removal apply when getting rid of brush-applied paint.

4.4 Ball-point pen

The composition and intensity of ball-point pen markings may make them difficult to remove from permeable surfaces, such as vinyl backings on bus seats. Solvent-based removal agents should not be used. Fortunately, ball-point graffiti are usually small.

Most of the marks may be erased by scrubbing with water and detergent, judicious use of abrasive graffiti cream or rubbing with a pencil eraser. Ball-point pens cannot easily mark many hard and smooth surfaces, so the surfaces are easily cleaned. Persistent marks on painted surfaces may best be obliterated by a coating of paint.

4.5 Lipstick and wax crayon

Lipstick and crayon are not widely used for graffiti, but they may leave ghosting on permeable plastic surfaces. Removal methods are similar to those employed for graffiti produced with ball-point pens.

4.6 Chalk

Chalk is the oldest substance for making graffiti. It is easily removed from most surfaces by wiping or scrubbing, followed by washing. Because of its impermanence, it is seldom used for graffiti now.

4.7 Pencil

Pencil marks are easily removed with an eraser. But scratching may remain if the surface is soft or the pencil lead hard, and the surface may need re-finishing.

4.8 Knives and other scratching implements

Clearly, ceramic tiles, glass, vitreous enamel and other hard materials resist scratching by all but the hardest implements, such as diamond and hard steel. But softer surfaces, including normal paint, wood, soft brick and other soft surfaces can be permanently marked, and surface re-finishing may be the only solution.

4.9 Self-adhesive stickers

Stickers provide a rapid means of spreading slogans, political and social messages and illicit commercial advertisements.

For good adhesion, they are usually placed on smooth or semi-smooth surfaces, and the adhesives are normally of the solvent-based synthetic type.

4.9.1 Removing self-adhesive stickers

Ease of removal of self-adhesive stickers varies with the surfaces that the stickers are on. They are often placed on the smooth glass of train, bus and other windows, on ceramic tiles, terrazzo and other smooth, hard masonry surfaces, and on melamine-panelled train and bus interiors. Removal with a sharp-bladed scraper and a suitable graffiti removal solvent is quite easy. It is often useful to abrade or score the surface of the label to permit good solvent penetration and allow a brief soaking-in time, before taking the sticker off with a scraper. Often the sticker comes away intact.

Coarse abrasive powders and steel wool, which are a temptation to use to get rid of stubborn stickers, should be avoided. They produce unnecessary dust and may damage the underlying surface.

After stickers have been removed, excess adhesive and solvent must be wiped off and the surface cleaned with plain water or detergent.

Solvent-based agents may permanently damage vinyl cladding, some painted surfaces and other smooth but soft or permeable materials. Applying the removal agent to a small test patch in an inconspicuous place will indicate whether it is suitable before starting to use it.

4.10 Pasted fly posters (bill posters)

Strictly speaking, fly posters are not graffiti, but the problems they present are closely related to those of graffiti, and they often have to be dealt with by the same people. Fly posters are frequently pasted on outside walls and hoardings at night or at weekends, by organizations and individuals spreading illicit publicity for such things as pop concerts. They are often put up in small localities and appear repeatedly. They may make the environment very unsightly (Plate 5), and the problem can be aggravated by competitive fly posting by other people vying for highly visible sites.

Most posters are paper with water-soluble paste as adhesive, so solvents are not needed to remove them. Scoring the paper, soaking with water or special poster remover, and scraping off with a scraper or stiff brush is all that is required. Regular removal of fly posters may reduce re-posting since it will diminish their advertising value, and may eventually lead to the site being abandoned. Over-posting with blank paper or legitimate posters only encourages more graffiti or fly posting.

Groups with commercial interests are often responsible for clandestine fly posting, so detection and prosecution may be the best deterrent.

5 Graffiti removal agents

There has been a great increase in the incidence of graffiti in recent years, and as a result many removal products have appeared on the market, some of them claiming to be the ultimate solution to the problem. Needless to say, no product is effective in all situations and some are better than others for cleansing specific types of graffiti.

5.1 Solvent-based graffiti removers

Despite variations in composition, appearance and claimed efficacy, most graffiti removers rely on the inclusion of solvents for paints or inks. The solvents may be present as single components or, more usually, as a mixture of co-solvents. Since these base solvents are normally thin and free-flowing, their viscosity is often increased by the inclusion of inert thickening agents which produce semi-mobile gels or creams. The formulation may also include abrasives, and sometimes surfactants to improve wetting or help deposit removal.

5.1.1 Solvent-based graffiti removal liquids

Solvent-based graffiti removal liquids may be packaged in the following forms:

- screw-top can or bottle. The agent is applied to the surface with a cloth;
- trigger-action spray bottle, usually plastic. The agent is distributed over the surface as a medium-coarse spray and worked in with a cloth or brush;
- aerosol can. The pressurized agent is distributed on the surface as a fine spray and worked in with a cloth or brush.

One advantage of screw-top cans and bottles is that they come in a variety of sizes. And tinplate cans are impermeable to solvent, virtually unbreakable and cheap.

Pressurized aerosol applicators for solvent delivery are not recommended because, apart from the prevailing environmental concern about such appliances, the solvent which is frequently harmful in itself, is emitted in a finely dispersed form, increasing the possibility of ingestion or inhalation, particularly during close work or in confined spaces. There are further disadvantages:

- with some formulations, there is a danger of explosion of the pressurized product and of increased flammability;
- the packaging is expensive;
- it is difficult to assess how much active product remains in the container.

In any case, application by aerosol is not significantly easier than by either of the other two methods. Pump- or trigger-spray applicators are more acceptable because their contents are not pressurized and the spray is normally very much coarser and remains airborne for only a short time. Further advantages are that:

- many of them have a safety lock on the nozzle to prevent accidental discharge;
- the containers are usually translucent plastic which allows the remaining solvent content to be checked. They are also relatively lightweight;
- solvent can be applied quickly to the surface as a thin film, reducing wastage and keeping solvent vapour to a minimum.

Pump- or trigger-spray applicators are liked by operators because of their ease of use. However, the plastic of which the containers are made must be chosen with care. There can be solvent loss through the container walls and embrittlement or softening of the container. Shelf life of the product may consequently be affected.

Solvent-based graffiti removal products vary considerably in their effectiveness. Products containing potent chlorinated solvents, which are also used in paint strippers, will vigorously attack aerosol paint graffiti, but because of their volatility, their use must be restricted to open air locations and appropriate safety precautions must be strictly followed.

Low viscosity liquids run down vertical surfaces too quickly and the active solvent may evaporate before it has any effect. Such products are best applied manually with swabs and used for small tasks.

Great care should be taken in the use of aggressive graffiti removal solvents on painted surfaces, as the underlying paint will be attacked if the contact time is prolonged. A trial should be carried out on an unobtrusive section before using a solvent cleaner on paintwork.

Graffiti removal solvents are most suitable for use on smooth non-permeable unpainted surfaces. The products should be applied directly to the surface either from a pump-action container or with a clean swab. The affected area should be completely covered and the cleaner vigorously worked in with swabs. Smears may have to be removed with more solvent, but the amount should be kept to a minimum and wiped off with a clean dry swab. Finally, it will probably be necessary to wash down the cleaned parts and the adjacent areas with water and detergent to get a uniform finish.

Surfaces that may be cleaned with solvents include untextured glass, vitreous and stove enamels, stainless steel and glazed ceramic tiles. Epoxy- or urethane-coated surfaces, perspex and poly-carbonate may also respond to graffiti removal solvents if they are used cautiously with minimum solvent contact time. Lightly textured and unpainted non-permeable and non-porous surfaces may need the solvent to be worked in harder with a bristle brush.

In all cases, there must be good ventilation, and health and safety regulations relating to the use of solvents must be followed.

Graffiti removal solvent composition

At present, solvent types commonly used for graffiti removal may be divided into five groups:

- Group 1   containing chlorinated hydrocarbons
- Group 2   containing monoglycol ethers and glycol acetates
- Group 3   containing diglycol ethers
- Group 4   containing polar solvents
- Group 5   containing miscellaneous solvents.

Group 1 products contain potent and volatile solvents, often with low occupational exposure limits. For that reason, their use should be restricted to **open air locations** and appropriate safety precautions must be observed.

Group 2 products, because of the extremely low occupational exposure limits for their solvent base, should be used **only in the open air**, under supervision after realistic continuous vapour monitoring trials and with strict safety precautions.

Occupational exposure limits for Group 3 products have not been established. But the solvents are typified by low vapour densities which should give low vapour concentrations in use. The products are likely, therefore, to present an insignificant health and safety hazard in non-aerosol form, but before use, trial vapour monitoring should be carried out.

Most of the products in Group 4 contain N-methyl-2-pyrrolidone as the polar component which has strong solvent characteristics. Little reliable health and safety data is available at present but such products in **non-aerosol** form may be acceptable for special graffiti removal applications (e.g. from paintwork) where more established solvent-based products are not totally successful. But vapour monitoring trials should be held before adopting these products for graffiti removal.

The chemical composition of products in Group 5 varies greatly. Their safety and suitability for specific uses can only be considered on a case-by-case basis.

5.1.2 Solvent-based graffiti removal gels

Graffiti removal gels consist of solvent products combined with thickening or gelling agents. The solvents fall into the same five groups as the liquid solvent-based graffiti removal agents. Because they may emit toxic vapours, many gels should be used only in the open air. The principal advantages of incorporating graffiti removal solvents into gels are that:

- gels adhere to vertical surfaces, ensuring prolonged contact with the graffiti;
- solvent evaporation is usually retarded, prolonging the presence of the active component;
- evaporation retardation reduces instantaneous solvent vapour levels, although the cumulative amount of vapour produced will be the same over a longer period of time;
- gels are easily and controllably applied by a bristle brush.

Because of the viscous nature of the gels, diffusion of the active solvent into the graffiti markings takes time. So a minimum of 30 minutes contact time must be allowed. The process may be speeded up by working in the gel with a stiff bristle brush or broom, particularly on surfaces with a pronounced texture or where aerosol spray paints have been heavily applied and allowed to 'age'.

The inclusion of an abrasive ingredient helps break down

stubborn graffiti marks, and the action of the gel can be reinforced if it is followed by the use of medium-pressure hot water jet cleaning.

Most graffiti gels contain aggressive solvents and need prolonged contact, and as such, they should not be used on painted, permeable plastic or soft porous surfaces. They are most suitable for large, smooth to semi-rough expanses and exterior surfaces in open air locations where solvent vapours can disperse quickly and pressure water jet cleaning is possible.

5.1.3 Solvent-based graffiti removal creams

Graffiti removal creams are usually opaque substances that are applied manually with swabs. They are solvent based with the addition of inert thickeners and a mild abrasive. Their main advantages are:

- they can be used for graffiti removal on interior surfaces, e.g. inside train carriages and other vehicles;
- they are easily applied directly from the container and spread manually with swabs;
- the consistency of the cream reduces solvent evaporation and increases active contact time;
- the thicker formulation also reduces instantaneous vapour levels (but there must be adequate ventilation);
- the presence of a mild abrasive may help in the removal of stubborn graffiti, although repeated use may cause slight deterioration of some surfaces.

The cream may have to be worked into textured surfaces, and be removed with a bristle brush, but it can generally be wiped off easily with a clean, dry swab. To ensure that all the product is removed and a chalky residue of abrasive matter is not left, the cleaned area and surrounding surface should be washed down with a minimal quantity of detergent in water.

The cream may also be used where a cleaning solvent has been applied and has left ghosting on melamine or painted surfaces.

Chemical and abrasive additives

Abrasive and chemical additives are often included in gel, cream and other formulations. Indeed, many pressure washers have a facility for introducing detergents and other chemical agents or abrasive grit such as sand into the water jet, but the combination of pressure washing and such additives is not for general use. It could interfere with the action of previously applied graffiti remover, and chemicals under pressure may be driven deep into the material and cause problems later on if it is decided to add a protective coating. In addition, grit may damage the surface. Non-porous surfaces may, however, benefit from a general cleaning and washing down with chemical additives.

5.2 Bleach-based graffiti removers

Some agents for the removal of certain graffiti, such as felt-tip pen marks, incorporate bleaches which destroy rather than dissolve the colour. Bleaches, like solvents, need time to work, so an inert filler is normally embodied in the product to produce a mortar or a gel that spreads easily and adheres to both vertical and overhead surfaces, ensuring adequate contact time. Some products work better because of a 'poultice' action on felt-tip graffiti on porous and permeable surfaces.

A graffiti removal bleach can remove dye-based felt-tip or other inks from permeable and porous surfaces, but it will be no use against aerosol spray paints where the polymer content prevents effective bleaching. It could well be applied, however, where there is a mixture of types of graffiti mark, before using solvent-based agents.

Graffiti removal bleaches have been used successfully to remove felt-tip pen graffiti from terrazzo, porous brick, concrete, light-coloured emulsion-painted render, rubber tiles and linoleum, but care should be taken both in the choice of the removal agent and the method of application. If possible, small scale tests should be done in unobtrusive areas before using bleaches on the main markings.

The mortar form must be left in contact with the surface long enough to do its job, normally about two hours. It is easily spread without splashing and is not flammable, but there must be sufficient ventilation to dispel the chlorine gas that hypochlorite bleaches are liable to give off. Skin and eyes must be protected during mixing and use.

Spent mortar or gel should be removed with a wide bladed scraper, put into a waste container and the surface should be washed with a detergent solution.

The mortar or gel has a limited shelf life, so it should be freshly made up for each application.

6 Graffiti removal equipment

6.1 Manual graffiti removal equipment

The manual removal of graffiti makes use of simple, easily available and inexpensive equipment such as is used for ordinary cleaning. But when chemical removal agents are used, suitable protective clothing and equipment must be provided, as well as facilities for the storage and disposal of chemicals, and there must be access to a water supply.

The basic range of equipment comprises swabs, scrapers, scourers, scrubbing brushes, mops, buckets and approved removal and general cleaning agents (Plate 6).

6.1.1 Swabs

A good supply of clean cotton swabs should always be available for both applying and cleaning off graffiti removal products, and for washing down after graffiti have been removed. The swabs should be free of dirt, grit, lint and fugitive dyestuff. They should have enough absorbency for the cleaning product and be tough enough to withstand vigorous rubbing in and wiping on moderately rough surfaces. Recycled rags should not be used as they vary in quality.

Soiled swabs should be put in metal bins with lids which, because of the danger from solvent vapours, should be marked with hazard warnings. After completion of a graffiti removal operation, the swabs should be taken from the site and appropriately disposed of.

6.1.2 Scrapers

Metal bladed scrapers are useful for removing thickly-painted graffiti and adhesive stickers before using solvent-based products to get rid of residual marks. They minimize the spreading and smearing of paint that can occur during cleaning, and they reduce the amount of graffiti remover needed. Scrapers with sharp replaceable blades are particularly effective on smooth surfaces such as glass and ceramic tiles. Fixed-blade paint scrapers, either flat-edged or profiled, are also useful, but care must be taken to see that they do not gouge or scratch the underlying surface.

Stiff plastic scrapers are better than metal ones for removing thick paint from plastic and other smooth, soft surfaces.

6.1.3 Scourers

Plastic mesh and other mild scouring pads may be used to remove stubborn graffiti from smooth or semi-smooth surfaces. They should be cleaned regularly during use to get rid of paint and debris. Steel wool is not generally suitable as it may leave noticeable scratches. Pumice and other scouring powders may also abrade surfaces. They should be used only as a last resort.

6.1.4 Scrubbing brushes

Small hand scrubbing brushes are useful for working the graffiti removal agent in, particularly when the surface is uneven or of a rough texture. Brushes with medium-stiff natural bristles are the most suitable. Synthetic bristle brushes should not be used as solvent-based graffiti removers may harm them. Steel or very harsh bristle brushes should also be avoided as they are likely to damage the surfaces they are applied to.

Brushes should be cleaned often during use to prevent the unnecessary spread of colourants, and they should be left clean after completion of the work.

6.1.5 Trigger-spray applicators

Trigger-spray applicators are the best devices for applying solvent-based graffiti removal products directly onto a surface. To minimize hazards from the product, containers should not exceed one-litre in capacity, and they should be filled under safe conditions away from the site. Suppliers often provide applicators ready-filled with product.

Applicators must be made of solvent-resistant materials, and the nozzles must have effective safety locks. All product containers must be properly labelled with the product name and the appropriate health and safety information.

6.2 Mechanical graffiti removal equipment

Most mechanical equipment used for graffiti removal is electrically powered, so its use at a particular location is dependent in the first instance on the availability of an appropriate power source. Some machinery may also need other services, such as mains water for pressure washing or compressed air of suitable pressure and quality for abrasive blasting or surface peening. If such facilities are not available on site, it may be feasible to use mobile generators or air compressors.

Good health and safety practice must be followed wherever electrical equipment is employed. The equipment should be connected to the mains through a suitable isolator and adequate cut-out switches or pressure relief valves must be installed. In some places, spark-proof equipment may be necessary. Electrical appliances must be operated only by properly trained and experienced staff, and appropriate protective clothing and equipment must be worn (Plate 7). Particular care must be taken at any location where electrical equipment is used with water, for example with pressure washers, and especially in the vicinity of electrified track.

In locations where smoke and excessive noise are unacceptable, restrictions should be placed on the use of diesel-powered equipment. And in some areas, restricted physical access may preclude the use of large scale equipment.

By far the most widely used graffiti removal machines are pressure washers of various types. Hand-held abrasive tools and grit blasting equipment should be used sparingly as they may badly damage the surfaces being cleaned.

6.2.1 Pressure washers

The use of medium to high pressure water washing equipment, together with manually applied chemical removal agents, is a common way of tackling graffiti. It is, at present, the most effective means of removing graffiti from large open air expanses and the exteriors of buildings. For the best results, the correct procedures and machine operating conditions must be chosen and applied.

A wide variety of machines for delivering water under pressure is now available – these may be called pressure washers, power washers or jet washers. They come in a range of sizes and offer an assortment of facilities and performance features. Five main factors, which will be discussed in turn, influence the effectiveness of pressure washing operations. These are:

1. pressure rating;
2. water flow rate;
3. spray nozzle design (type, size and angle);
4. water temperature;
5. chemical or abrasive additives introduced into the stream.

1. Pressure rating

Pressure rating is the most emphasized selling feature in equipment descriptions.

(a) *Low pressure (less than 100 psi)* is used only for 'water mist' conservation cleaning of delicate masonry.

(b) *Medium to high pressure (500 to 5000 psi)* is used for all general purpose cleaning and graffiti removal. In practice, most graffiti removal water jetting equipment (Plate 8) operates in the 1500 to 3000 psi range, depending on the type of surface to be cleaned. Most major manufacturers produce a series of machines in this range with different performance specifications.

(c) *Ultra-high pressure (10,000 psi and above)* is employed only for 'hydroblasting', an extremely severe process for heavy duty work such as boiler descaling or stripping off concrete surfaces, rather than graffiti removal.

Pressure rating is determined at the nozzle and, clearly, alters with distance from the surface being sprayed.

2. Water flow rate

The rate of supply of water to the sprayhead determines the impact force of the spray just as much as the pressure rating will. For soft masonry surfaces, a high flow rate at reduced pressure may be the best option because a good volume of water, flushing away dirt, cleaner and graffiti residues, decreases the risk of erosion of the surface.

On the other hand, in locations where containment and a minimal use of water are important, low flow rates (down to 3 gal/min) with high pressure may be a better combination. The greatest efficiency is likely to be achieved with water flow rates of between 4 to 8 gal/min.

3. Spray nozzle design (Fig. 6.1)

Spray nozzles on pressure washer lances are usually interchangeable and may have a 'roll over' facility for a quick change from one nozzle type to another.

Nozzle design determines the shape of the spray and has a strong bearing on the success of the cleaning process. A fan-type nozzle providing a spray fan of an angle of 15–50° is considered best for graffiti removal and general surface cleaning. Larger angles reduce spray impact but increase area coverage, while a 0° pencil jet produces an intense impact harmful to soft or crumbly masonry surfaces but useful for removing graffiti from hard surfaces that have resisted more gentle fan jets. Hollow cone sprays are no advantage for most surfaces, and may damage those that are soft or friable.

4. Water temperature

Many pressure washers have a facility for heating the water, usually by circulation past a diesel-fuelled burner, up to a temperature of 100°C. Heated water jetting may be useful for graffiti removal from metal surfaces where expansion of the metal helps break the bond, and in cold weather when the heat may speed up chemical reactivity and soften paint or grease deposits.

Heated water, however, if used in combination with solvents, may cause flash evaporation of the solvents, reducing their graffiti removal effectiveness. Machines with heating facilities can usually be run as cold water washers as well, and trials with both hot and cold water will show which is the better for the task.

In some cases, leaving graffiti removal agents in contact with surfaces for longer than usual before washing them off, is more effective than using hot water. In addition, at some locations, diesel-fuelled heating may be unacceptable because of the risks of fumes and fire.

5. Chemical and abrasive additives

Incorporating chemical or abrasive additives into the water is not recommended except on selected hard non-porous surfaces. As with

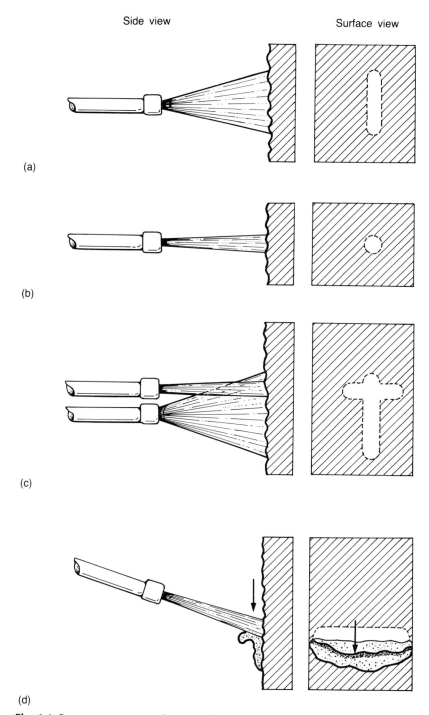

Fig. 6.1 Some pressure washer nozzle designs: (a) fan nozzle; (b) pencil nozzle; (c) cross fan (twin) nozzle; (d) removal action of fan nozzle.

other pressure cleaning techniques, the effectiveness of the work depends largely on the skills of the operators. Only qualified, trained staff should be used for pressure washing.

As a result of the potential dangers of both graffiti removal agents and high water pressures, protective clothing and equipment must be worn by everyone involved in pressure cleaning work, and run-off water must be properly contained.

6.2.2 Abrasive blasting

Abrasive blasting machines, employing a wide range of abrasive media and various methods of delivery, are widely available. Conventional abrasive media, such as sand and pulverized slag, are not generally suitable for graffiti removal because they are likely to damage surfaces. Furthermore, open-jet blasting is unsuitable for use in interior locations because it produces large quantities of dust.

A modified technique, wet grit blasting, which incorporates the grit in water, tends to reduce the dust hazard. Similarly, lower pressure closed-cup techniques, in which the abrasive and surface debris are captured and recycled, keep down the dust, but the operation is slow and getting an even finish with such methods is difficult.

Alternative abrasive media, such as friable plastics, are becoming available in modern closed-cup abrasive blast machines. They offer the advantage of selectively removing paint without damaging the surface but the process is slow and not generally suited to graffiti removal.

6.2.3 Surface peening

Low pressure blasting equipment, using small glass beads to peen certain ductile surfaces, usually metal, is available. The beads remove very thin porous surface oxide coatings caused by weathering, and they close pores on the surface of the metal.

This method has effectively removed graffiti, in small scale trials, from weathered aluminium surfaces and renovated the weathered metal to a good finish. But the process is slow and demands skilled operation. Faster open jetting could be done only in a large blasting chamber and operators would have to wear protective suits and respirators. The cost of the work and the glass beads is relatively high.

7 Graffiti removal methods

7.1 General aspects

Graffiti removal is essentially a cleaning operation, even if it is often difficult and of a specialized nature. It is work that employs the methods, cleansing agents and equipment used in other cleaning activities. It may be additional to routine or renovation cleaning, but it cannot be isolated from it.

There is little point in removing graffiti from part of a soiled surface without cleaning the whole surface to the same standard. Although that may seem obvious, it is not unusual for a space from which the graffiti have been removed to stand out because it is cleaner than the surrounding surface. In addition, often, graffiti that could be removed without difficulty are imperfectly erased, leaving unsightly smear marks either because the cleaning method was wrong or the work was not finished off properly. The results, in such cases, are not only unacceptable, but may encourage graffitists, perceiving a failure on the part of the 'enemy', to re-attack the areas.

The aims of graffiti removal and general cleaning are the same – to remove contaminant without spoiling or altering the original surface. To achieve this, graffiti removal (and cleaning) should be carried out as gently as possible, consistent with achieving satisfactory results and acceptable costs, safety and speed of working.

There are two basic approaches to graffiti removal – manual and mechanical. In some instances, a combination of the two may be needed.

7.2 Manual graffiti removal methods

Manual methods include wiping with a swab, scraping, scouring with abrasive papers or pads and washing down. These methods normally entail the use of chemical graffiti removal agents.

The main advantages of manual methods are:

- adaptability to a variety of locations, including those that are remote or in restricted spaces, such as carriage or bus interiors, where mechanical methods are not suitable and mains services such as electricity are not required;
- protection of surrounding surfaces. Operators experienced in graffiti removal will select the right cleaning methods for particular surfaces, such as melamine or glass in vehicle interiors, will avoid touching neighbouring surfaces and will not need to use surface masking;
- safety. No electric- or diesel-powered tools are used, and little water is required for washing down cleaned surfaces, so operational risks are reduced;
- low cost because no sophisticated equipment is used;
- ease of deployment of mobile flying squads to respond quickly to graffiti attacks.

The disadvantages of manual methods are:

- slowness compared with mechanical methods;
- labour costs. Manual methods tend to be labour intensive;
- unpopularity of the work because of the need to wear protective clothing when working with chemical agents;
- hazards of chemical agents. When solvent-based graffiti removal agents have to be used, exposure of operators to harmful vapours during close work must be minimized. This means using only carefully chosen removal agents, ensuring good ventilation (so restricting locations) and limiting exposure time (reducing active work time).

7.3 Mechanical graffiti removal methods

The most commonly used mechanical methods include the following:

(a) hot or cold pressure water jet cleaning, with or without chemical additives;

(b) pressurized air or water grit blasting, using any of a variety of abrading media. The abrasive medium may be recovered by vacuum;

(c) linishing (belt, disc or orbital abrasion).

7.3.1 Water jet cleaning (pressure washing)

This is usually done after graffiti remover has been applied. Medium to high pressure (1500–3000 psi) water jetting may help the graffiti remover do its work and it removes residues and rinses the surface. Abrasives are sometimes incorporated in the water which may be hot (100°C) or cold. The process is most suitable for exterior surfaces in locations where there is good drainage.

The operation is fairly fast, and the labour and equipment costs are moderate.

The advantages of water jet cleaning are that it can be adapted to suit the surface type and the nature of the graffiti, it is suitable for large areas, it will not damage most surfaces, particularly if abrasives are not used, and the equipment is reasonably mobile. Some pressure washers now exist with their own water supplies.

The main disadvantage of water jet cleaning is that it is not suitable for most enclosed locations unless there is good ventilation and drainage.

7.3.2 Pressurized grit blasting

Blasting with high pressure air or water containing an abrasive material will remove most graffiti, and also the surface layer. To minimize damage to the surface, less abrasive media of varying degrees of hardness are now available. This system is widely used for cleaning building exteriors.

The process tends to be slower than water jetting and the cost is higher because some of the abrasive substance will be lost, even with vacuum recovery.

The advantages of the method are that it is thorough, no chemical agents are required and, with air blasting, no water or drainage is needed.

The disadvantages are: dust which might present a health and environmental hazard, the likelihood of surface damage to soft or glazed surfaces, the possibility of drain blockage when wet blasting, and the bulkiness of the equipment which may limit access.

Grit blasting is a rather aggressive means of removing graffiti and has a tendency to erode soft brick, pointing and other surfaces, so it should be resorted to only where more gentle methods are impracticable or have failed, or where the surface is tough enough to withstand the process.

7.3.3 Linishing/abrading

Linishing is a method of mechanical abrasion through the use of disc, belt or orbital sanders. It is used widely for small scale work on certain surfaces and removes the top surface as well as the graffiti.

It is a rather slow process and can be expensive in labour costs.

Its advantages are that the work needs no chemical agents, water or drainage (although some water may be needed for cleaning up after the work is done) and the equipment is cheap and easily portable.

Its disadvantages are that the method may scar or change the appearance of the surface, it creates dust, and manual finishing may be required.

7.4 Costs of equipment

The cost of mechanical equipment for a specific graffiti removal task may be high, but such equipment is often available for hire. Alternatively, some private companies specialize in cleaning and graffiti removal. Before engaging a company to do the work, its expertise, cleaning methods and removal media should be investigated and small scale site trials should be carried out.

8 Selection guide to graffiti removal methods

8.1 How to use the selection guide

The selection guide to graffiti removal methods (GRMs) specifies the best methods so far devised for cleaning graffiti from particular surfaces. Each GRM is set out in detail in the section following this selection guide.

GRMs specify the materials, equipment and methods needed for erasure of the markings. They also give health and safety advice and details of any other necessary precautions.

To use the guide:

1. identify the surface-type to be cleaned (column 1);
2. identify the marker type(s) that produced the graffiti (column 2);
3. note the GRM number (column 3);
4. turn to that GRM in the next section.

Selection guide to graffiti removal methods

Table 8.1 Selection guide to graffiti removal methods

Surface type	Graffiti marker type	Graffiti removal method	Notes
Glazing			
Clear glass – windows, mirrors	Aerosol paint ⎫ Felt-tip ⎬ Wax crayon ⎭	GRM 2	
	Adhesive stickers	GRM 7	
'Perspex' (acrylics) Polycarbonate	Aerosol paint ⎫ Felt-tip ⎬ Wax crayon ⎭	GRM 8	(b) (c)
	Adhesive stickers	GRM 10	
Paintwork			
Gloss (oleoresinous, alkyd)	Aerosol paint ⎫ Felt-tip ⎬	GRM 8	(a) (c) (a) (c) (d)
	Adhesive stickers	GRM 10	
Urethane and epoxy paints, lacquers	Aerosol paint ⎫ Felt-tip ⎬	GRM 15	
	Adhesive stickers	GRM 10	
Stove enamel	Aerosol paint ⎫ Felt-tip ⎬	GRM 2	
	Adhesive stickers	GRM 7	
Emulsion paint (smooth and textured surfaces)	Aerosol paint	GRM 8	(a) (c) (d)
	Felt-tip	GRM 3	(b)
	Adhesive stickers	GRM 10	
Exterior textured paints	Aerosol paint	GRM 6	
	Felt-tip	GRM 5	
Rubber			
	Aerosol paint ⎫ Felt-tip ⎬	GRM 8	(c)
Rubber flooring/floor tiles	Aerosol paint	GRM 8	(c)
	Felt-tip	GRM 3	
Metals			
Aluminium (weathered, unpainted)	Aerosol paint ⎫ Felt-tip ⎬ Wax crayon ⎭	GRM 1	
	Adhesive stickers	GRM 10	
Aluminium (anodized)	Aerosol paint ⎫ Felt-tip ⎬ Wax crayon ⎭	GRM 2	
	Adhesive stickers	GRM 10	

Table 8.1 *Continued*

Surface type	Graffiti marker type	Graffiti removal method	Notes
Metals *continued*			
Steel (galvanized, unpainted)	Aerosol paint ⎫ Felt-tip ⎭	GRM 1	
	Fly posters	GRM 13	
Vinyl (PVC)			
	Aerosol paints ⎫ Felt-tip ⎭	–	(d)
	Adhesive stickers	GRM 10	
Laminates			
Melamines (Formica, Warerite, Polyray)	Aerosol paint ⎫ Felt-tip ⎭	GRM 9	
	Adhesive stickers	GRM 10	
'Melaminium' (aluminium backing)	Aerosol paint ⎫ Felt-tip ⎭	GRM 9	
	Adhesive stickers	GRM 10	
GRP (glass reinforced plastic)			
	Aerosol paint ⎫ Felt-tip ⎭	GRM 8	
	Adhesive stickers	GRM 10	
Vitreous materials			
Ceramic tiles (hard glaze)	Aerosol paint	GRM 2	(c)
Vitreous enamel	Felt-tip ⎫ Wax crayon ⎭	GRM 3	(c)
	Adhesive stickers	GRM 7	
Woodwork (unpainted	Aerosol paint ⎫ Felt-tip ⎬ Scratches ⎭	–	(e)
Masonry (Interior surfaces)			
Brick – porous (London Yellow, Common Stock)	Aerosol paint ⎫ Wax crayon ⎭	GRM 4	
	Felt-tip	GRM 3	
Brick – glazed (semi-engineering facing)	Aerosol paint ⎫ Wax crayon ⎭	GRM 4	
	Felt-tip	GRM 3	

Selection guide to graffiti removal methods

Table 8.1 *Continued*

Surface type	Graffiti marker type	Graffiti removal method	Notes
Masonry (Interior surfaces) *continued*			
Grouting (cementitious)	Aerosol paint ⎫ Wax crayon ⎬ Felt-tip ⎭	–	(f)
Grouting (epoxy resin-based)	Aerosol paint ⎫ Wax crayon ⎬ Felt-tip ⎭	–	
Marble – Travertine (unfilled), resin-bonded, mosaic	Aerosol paint ⎫ Wax crayon ⎭	GRM 4	
	Felt-tip	GRM 3	
	Adhesive stickers	GRM 7	
Granite (smooth)	Aerosol paint ⎫ Wax crayon ⎭	GRM 4	
	Felt-tip	GRM 3	
	Adhesive stickers	GRM 7	
Concrete (smooth)	Aerosol paint ⎫ Wax crayon ⎭	GRM 4	
	Felt-tip	GRM 3	
	Chalk	GRM 14	
Terrazzo (walls and floors)	Aerosol paint ⎫ Wax crayon ⎭	GRM 4	
	Felt-tip	GRM 3	
	Adhesive stickers	GRM 7	
	Ball-point pen	GRM 12	
Cement render	Aerosol paint ⎫ Wax crayon ⎭	GRM 4	
	Felt-tip	GRM 3	
Masonry (Exterior surfaces)			
Concrete (smooth or rough)	Aerosol paint ⎫ Wax crayon ⎭	GRM 6	
	Felt-tip	GRM 5	
	Adhesive stickers	GRM 7	
	Fly posters	GRM 13	
Brick – porous (London Yellow, Common Stock)	Aerosol paint ⎫ Wax crayon ⎭	GRM 6	
	Felt-tip	GRM 5	
Brick – glazed (semi-engineering facing)	Aerosol paint ⎫ Wax crayon ⎭	GRM 6	
	Felt-tip	GRM 5	
Grouting (cementitious)	All types	–	(f)

Table 8.1 *Continued*

Surface type	Graffiti marker type	Graffiti removal method	Notes
Masonry (Exterior surfaces) *continued*			
Grouting (epoxy resin-based)	All types	–	
Granite (smooth) Portland stone	Aerosol paint ⎫ Wax crayon ⎬	GRM 6	
York stone	Felt-tip	GRM 5	
	Adhesive stickers	GRM 7	
	Chalk	GRM 14	
Mosaic marble	Aerosol paint ⎫ Wax crayon ⎬	GRM 6	
	Felt-tip	GRM 5	
	Adhesive stickers	GRM 7	
Cement render	Aerosol paint ⎫ Wax crayon ⎬	GRM 6	
	Felt-tip	GRM 5	
	Chalk	GRM 14	
Ceramic tile	Aerosol paint ⎫ Wax crayon ⎬ Felt-tip ⎭	GRM 2	
	Adhesive stickers	GRM 7	

Notes
(a) Original paintwork may need to be made good.
(b) Allow minimum solvent contact; avoid harsh scraping and abrasives.
(c) Try small test patch first.
(d) Regardless of GRM used, graffiti ghosts persist. Protect cladding with surface coating or replace with graffiti resistant paint or laminate.
(e) Sand down and re-finish.
(f) Graffiti marks are unlikely to be totally removed. If necessary, wide cementitious grouts should be replaced with epoxy grouting or protected with anti-graffiti surface coating. Dark grouting is preferable to light.

GRM 1   Graffiti removal method one

Applicability

> *Surfaces:* Unpainted aluminium exteriors, galvanized steel
> *Graffiti types:* Permanent felt-tip pen, aerosol spray paints, wax crayon, lipstick

Materials and equipment

- Graffiti removal gel or similar agent
- Long-handled bristle brooms, clean cotton swabs
- Medium/high pressure heated water washer, hoses and lance with fan jet
- Gloves and eye protection, waterproofs and boots

Health and safety advice

- Graffiti removal gel can be **flammable** and **irritant**.
- Wear gloves and eye protection. Keep away from sources of ignition or direct heat.
- Ensure good ventilation and do not smoke.
- Wash splashes off skin with plenty of water. If splashed in eyes, wash immediately with plenty of water and seek medical advice.
- Do not use in enclosed areas.
- Only sufficient gel should be dispensed into working containers for immediate work needs.

Graffiti removal procedure (see Plates 9–12)

1. Apply gel by stiff brush to surface, working it well in.
2. Leave at least 30 minutes – longer if possible.
3. Spray off surface with pressure washer (70°C, 1500 psi).
4. Re-apply fresh gel to stubborn graffiti and leave as before.
5. Wipe with clean swab or brush down.
6. Wash down with warm water and detergent.

Limitations/precautions for surfaces

- Do not use gel on painted or plastic surfaces. Adjoining painted or plastic surfaces must be masked off before applying gel.
- Do not use abrasive grit in pressure machine.
- For stubborn graffiti, use graffiti gel containing mild abrasive and rinse down thoroughly after use.

GRM 2   Graffiti removal method two

Applicability

Surfaces: Clear, untextured glass, anodized aluminium, stainless steel, vitreous and stove enamels, hard glaze ceramic tiles

Graffiti types: Permanent felt-tip pen, aerosol spray paint, wax crayon

Materials and equipment

- Graffiti removal solvent
- Clean cotton swabs
- Metal disposal bin with lid
- Metal-bladed scrapers
- Gloves and eye protection

Health and safety advice

- Graffiti removal solvents can be **flammable** and **irritant** to eyes and skin.
- Wear gloves and eye protection.
- Ensure adequate ventilation and do not smoke.
- Place all used swabs in bin and remove from site for disposal.

Graffiti removal procedure

1. Remove thick paint carefully with scrapers.
2. Remove residual paint and felt-tip marks with solvent, wiping over thoroughly with clean swabs.
3. Remove solvent with clean dry swabs.
4. Wash down completely with swabs and water/water with detergent and allow to dry.
5. Clean windows with window polish, if necessary.

Limitations/precautions for surfaces

- Take care not to scratch the surface with scrapers.
- Do not use steel wool.
- Avoid prolonged contact by solvent with any adjacent painted surfaces.

GRM 3　Graffiti removal method three

Applicability

Surfaces: Smooth concrete and brick, terrazzo, smooth resin-bonded granite, rubber flooring, some paints and emulsions (pale shades)
Graffiti types: 'Permanent' felt-tip pen *only*

Materials and equipment

- Graffiti removal bleach with filler
- Mixing spatula
- Plastering trowels
- Stiff hand brushes
- Steel wide-bladed hand scrapers
- Clean cotton swabs, mops and bucket
- Gloves and eye protection

Health and safety advice

- The bleach (sodium hypochlorite) is **corrosive** and **causes burns**.
- Contact with acids liberates toxic gas.
- Wear gloves and eye protection.
- Ensure good ventilation.
- If the product gets into eyes, rinse with plenty of water and seek medical advice.
- Wash off skin splashes with plenty of water.

Graffiti removal procedure (see Plates 13–16)

1. Ensure surface is free of oil, grease or previously-applied sealant. Scrub down with detergent solution if necessary and allow to dry.
2. Mix bleach and filler with spatula to produce a mortar of smooth consistency.
3. Apply mortar to surface with trowel, working in well with a brush and ensuring complete coverage of the graffiti.
4. Leave mortar for *at least* two hours in contact with the surface.
5. Remove mortar with scraper, collecting all debris for disposal.
6. Re-apply fresh mortar to any stubborn graffiti and repeat steps 4 and 5.
7. Wash surface thoroughly with clean water and allow to dry completely.

Limitations/precautions for surfaces

■ This is the *only* effective method at present for porous/permeable surfaces with felt-tip marks.
■ Ensure bleach does not affect the colour of the surface. Try a test patch before starting to deal with the graffiti.

GRM 4   Graffiti removal method four

Applicability

> *Surfaces:* Smooth concrete and brick, terrazzo, smooth resin-bonded granite, rubber flooring, some paints and emulsions
> *Graffiti types:* Aerosol paints, wax crayon, lipstick

Materials and equipment

- Graffiti removal solvent
- Sharp steel hand scrapers
- Clean cotton swabs
- Metal swab disposal bin with lid
- Gloves and eye protection

Health and safety advice

- Graffiti removal solvents can be **flammable** and **irritant** to eyes and skin.
- Wear gloves and eye protection.
- Ensure adequate ventilation and do not smoke.
- Place all used swabs in bin and remove from site for disposal.

Graffiti removal procedure

1. Ensure surface is free of oil, grease, previously-applied sealant and felt-tip graffiti marks. Remove felt-tip marks by GRM 3. Otherwise, scrub the surface with detergent solution and allow to dry.
2. Remove any thick paint deposits carefully with scrapers.
3. Apply cleaning solvent on clean cotton swabs and wipe over thoroughly with clean dry swabs.
4. Wash down the whole surface with water or detergent solution and allow to dry completely.

Limitations/precautions for surfaces

- Take care not to scratch the surface with scrapers.
- Use the method only after removing felt-tip graffiti by GRM 3.
- Do not use steel wool or harsh abrasives. Use a bristle brush to work in the solvent on textured surfaces.
- Avoid prolonged contact by solvent with any adjacent painted surfaces.

GRM 5 Graffiti removal method five

Applicability

Surfaces: Rough concrete, smooth or rough soft/semi-glazed brick (unpainted)
Graffiti types: 'Permanent' felt-tip pen *only*

Materials and equipment

- Graffiti removal bleach with filler
- Mixing spatula and plastering trowel
- Stiff hand brushes
- Medium/high pressure (1500–2000 psi) cold water washer with hoses, lance and fan jet
- Bristle brushes, mops and bucket
- Gloves and eye protection, waterproofs and boots

Health and safety advice

- The bleach is **corrosive** and **causes burns.**
- Contact with acid liberates toxic gas.
- Wear gloves and eye protection.
- Ensure good ventilation.
- If the product gets into eyes, rinse with plenty of water and seek medical advice. Wash off skin splashes with plenty of water.
- Permit only trained staff to use pressure washers.

Graffiti removal procedure

1. Ensure surface is free of oil, grease or previously-applied sealant. Scrub down with detergent solution if necessary and allow to dry.
2. Mix bleach and filler with spatula to produce a mortar of smooth consistency.
3. Apply mortar to surface by trowel. (Work in with brush on rough surfaces.) Ensure the graffiti are completely covered.
4. Leave mortar for *at least* two hours in contact with the surface.
5. Remove mortar by cold water jetting.
6. Re-apply fresh mortar to any stubborn graffiti and repeat steps 4 and 5.
7. Wash surface thoroughly with water and allow to dry completely.

Limitations/precautions for surfaces

- This method is *only* for felt-tip graffiti on porous surfaces. GRM 6 is appropriate for other markings.
- Ensure the bleach does not affect colour or surface. Try a test patch first.
- Do not use abrasives in water jetting.

GRM 6 Graffiti removal method six

Applicability

> *Surfaces:* Rough concrete, smooth or rough soft/semi-glazed brick
> *Graffiti types:* Aerosol paint, wax crayon, lipstick

Materials and equipment

- Graffiti removal gel or similar agent
- Long-handled bristle brooms, clean cotton swabs
- Medium/high pressure water washer, hoses and lance with fan jet
- Gloves and eye protection, waterproofs and boots

Health and safety advice

- Graffiti removal gel can be **flammable** and **irritant**. It is harmful if inhaled and can injure the skin.
- Wear gloves and eye protection.
- Keep away from sources of ignition and direct heat.
- Ensure good ventilation and do not smoke.
- Wash splashes off skin with plenty of water.
- If splashed in eyes, wash immediately with plenty of water and seek medical advice.
- Do not use in enclosed areas.
- Permit only trained staff to use pressure washers.
- Only sufficient gel should be dispensed into working containers for immediate work needs.

Graffiti removal procedure

1. Apply gel by stiff brush to surface, working it well in.
2. Leave at least two hours (longer if possible).
3. Spray off surface with pressure washer (hot or cold water, 1500–2000 psi).
4. Re-apply fresh gel to stubborn graffiti and leave as before.
5. Wash down with pressure water spray using clean water or detergent solution.

Limitations/precautions for surfaces

- Do not use gel on painted or plastic surfaces.
- Do not use abrasive grit in pressure washing machine.
- For stubborn graffiti, use graffiti gel containing mild abrasive and rinse down thoroughly afterwards.

GRM 7   Graffiti removal method seven

Applicability

Surfaces: Hardglaze ceramic tiles, clear untextured glass, stove enamel, vitreous enamel, terrazzo
Graffiti types: Self-adhesive stickers

Materials and equipment

- Graffiti removal solvent
- Supply of clean cotton swabs
- Plastic scourers
- Sharp metal-bladed hand scrapers
- Gloves and eye protection
- Metal disposal bin with lid

Health and safety advice

Graffiti removal solvents can be **flammable** and **irritant** to eyes and skin.
- Wear gloves and eye protection.
- Ensure adequate ventilation and do not smoke.
- Place all used swabs in the bin and remove from site for disposal.

Graffiti removal procedure

1. Lightly abrade the surface of the adhesive stickers with a scourer to allow the solvent to penetrate.
2. Apply cleaning solvent to stickers with swabs and allow to soak in (5–10 minutes).
3. Remove stickers with scraper, starting at outer edges.
4. Remove residual adhesive with solvent and swabs or scraper if necessary.
5. Wash down the surface with detergent solution and polish dry with clean swabs.

Limitations/precautions for surfaces

- This method is not for use on painted, plastic or soft surfaces.
- Glass (windows and partitions) may subsequently be polished with window cleaner and clean cotton swabs.
- Avoid prolonged contact by solvent with any adjacent painted surfaces.

GRM 8　Graffiti removal method eight

Applicability

> *Surfaces:* 'Perspex', polycarbonate, 'Melinex', GRP, oleoresinous ('gloss') paints, linoleum
> *Graffiti types:* Permanent felt-tip pen, aerosol paints, wax crayon, lipstick

Materials and equipment

- Graffiti removal solvent
- Clean cotton swabs
- Metal disposal bin with lid
- Gloves and eye protection

Health and safety advice

- Graffiti removal solvents can be **flammable** and **irritant** to eyes and skin.
- Wear gloves and eye protection.
- Ensure adequate ventilation and do not smoke.
- Place all used swabs in bin and remove from site for disposal.

Graffiti removal procedure

1. Apply graffiti removal solvent to surface with clean cotton swabs.
2. Allow the solvent minimum possible contact time before removing it from the surface with clean dry swabs.
3. Wash the surface with clean swabs and water or detergent solution.
4. Polish dry with clean swabs.
5. Use caution and minimum hand pressure throughout to avoid scratches and damage by solvent.

Limitations/precautions for surfaces

- Plastic glazing ('Perspex', 'Melinex', polycarbonate) may be finally polished with non-abrasive window cleaner and clean cotton swabs.
- Try a small test area for method suitability first.
- Do not use abrasive pads or powders.

GRM 9   Graffiti removal method nine

Applicability

Surfaces: 'Melamine'-faced panels, 'Melaminium'
Graffiti types: Permanent felt-tip marker, aerosol spray paint, wax crayon, lipstick

Materials and equipment

- Graffiti removal solvent or cream
- Clean cotton swabs
- Medium/soft bristle brushes
- Metal disposal bin with lid
- Gloves and eye protection

Health and safety advice

- Graffiti removal solvents and creams are **flammable** and **irritant** to eyes and skin.
- Wear gloves and eye protection.
- Ensure good ventilation and do not smoke.
- Place all used swabs in the bin and remove from site for disposal.

Graffiti removal procedure

1. Apply the solvent and work it well in.
2. Leave it on the surface for up to 5 minutes.
3. Remove the solvent with clean swabs (for smooth surfaces) or brush and swab (for textured surfaces).
4. Wipe the surface with swabs dampened with detergent solution. Use a brush for textured surfaces.
5. Wipe with damp clean swabs.
6. Put all used swabs in the bin and remove from the site.

Limitations/precautions for surfaces

- Remove residual 'ghosting' marks with graffiti removal cream applied by swab and wiped off with clean dry swabs, followed by a wipe down with water or a detergent solution.
- Do not use wire wool, pumice or other harsh abrasives.

GRM 10   Graffiti removal method ten

Applicability

> *Surfaces:* Anodized aluminium, stainless steel, 'Melamines', some plastics (the method is not suitable for some solvent-sensitive plastics, e.g. vinyls or paints)
> *Graffiti types:* Self-adhesive stickers

Materials and equipment

- Graffiti removal solvent
- Clean cotton swabs
- Plastic scourers
- Plastic-bladed scrapers
- Metal disposal bin with lid
- Gloves and eye protection

Health and safety advice

- Graffiti removal solvents can be **flammable** and **irritant** to skin and eyes.
- Wear gloves and eye protection.
- Ensure adequate ventilation and do not smoke.
- Place all used swabs in the bin and remove from the site for disposal.

Graffiti removal procedure

1. Lightly abrade the surface of the stickers with a scourer to allow the solvent to penetrate.
2. Apply solvent to stickers with swabs and allow to soak in for 5 minutes. (*Caution:* some plastics may be softened.)
3. Remove stickers with plastic scrapers, starting at outer edges.
4. Remove residual adhesive with solvent, plastic scourer and swabs, as necessary.
5. Wash down the surface with detergent solution, rinse with clean water and wipe dry with clean swabs.

Limitations/precautions for surfaces

- Do not use steel wool, metal scourers or metal scrapers. They will scratch the surface.
- Avoid prolonged contact by solvent with any adjacent painted surfaces.

GRM 11   Graffiti removal method eleven

Applicability

> *Surfaces:* Terrazzo, 'Melamines', painted surfaces
> *Graffiti type:* Pencil

Materials and equipment

- Soft pencil erasers
- Clean cotton swabs
- Water and detergent

Health and safety advice

No special precautions necessary.

Graffiti removal procedure

1. Rub off all pencil graffiti with eraser, removing all smear marks.
2. Wash the surface with detergent solution and swabs, rinse with clean water and wipe dry.

Limitations/precautions for surfaces

- Hard pencil may cause permanent scratching in soft surfaces. If so, expert advice may be needed for surface refinishing.

Graffiti removal method twelve

Applicability

Surfaces: Terrazzo, smooth concrete, painted surfaces
Graffiti types: Ball-point pen

Materials and equipment

- Soft rubber pencil erasers
- Graffiti removal solvent
- Clean cotton swabs
- Metal disposal bin with lid
- Gloves and eye protection

Health and safety advice

- Graffiti removal solvents can be **flammable** and **irritant** to skin and eyes.
- Wear gloves and eye protection.
- Ensure adequate ventilation and do not smoke.
- Place all used swabs in the bin and remove from the site for disposal.

Graffiti removal procedure

1. Rub off the bulk of the ball-point markings with the soft pencil eraser.

2. Wipe off the residual marks and smears with solvent and swabs, keeping contact time to a minimum.
3. Wash the surface with detergent solution and swabs, rinse with clean water and wipe dry.

Limitations/precautions for surfaces

- Intense ball-point pen marks may stain the surface if solvent is applied unless the bulk is previously removed.
- Test a small area first to check the suitability of the method.

Graffiti removal method thirteen   GRM 13

Applicability

Surfaces: Concrete, terrazzo, galvanized steel
Graffiti type: Fly posters

Materials and equipment

- Bill poster paste remover or detergent solution
- Stiff bristle brushes
- Medium pressure cold/hot water washer with hoses, lance and fan jet or mops, brooms and buckets
- Eye protection, waterproofs and boots

Health and safety advice

- Pressure water washers should be operated only by staff trained in their use, and eye protection and protective clothing must be worn.
- Beware of wet, slippery floors and pavements.

Graffiti removal procedure

1. Pre-soak posters with paste remover or detergent solution and scrub well in with brushes.
2. Leave for a few minutes.
3. Remove posters with high pressure jet or by scrubbing with brushes.
4. Wash down surrounding area, rinse and allow to dry.

Limitations/precautions for surfaces

- Do not use abrasive grit in pressure washers.

GRM 14　Graffiti removal method fourteen

Application

> *Surfaces:*　All
> *Graffiti type:*　Chalk (non-wax)

Materials and equipment

- Stiff bristle brushes
- Clean cotton swabs
- Water and detergent solution

Health and safety advice

- No special precautions necessary.

Graffiti removal procedure

Smooth surfaces (enamels, ceramic tiles, etc.)
Wipe off with swabs dampened with water or detergent solution.
Rinse with water and dry with clean swabs.

Rough surfaces
Remove bulk chalk marks with stiff bristle brushes and residues by
scrubbing with detergent solution. Rinse and allow to dry.

Limitations/precautions for surfaces

- No special limitations or precautions.

Graffiti removal method fifteen

GRM 15

Applicability

> *Surfaces:* 2 Pack polyurethane paint
> *Graffiti types:* Permanent felt-tip pen, aerosol spray paints

Materials and equipment

- Graffiti removal gel (not containing chlorinated hydrocarbon solvents)
- Soft bristle brushes
- Hose with hot or cold water. For large areas medium/high pressure heated water washer, hoses and fan jet
- Gloves and eye protection, waterproofs and boots

Health and safety advice

- Graffiti removal gel can be **flammable** and **irritant** to skin and eyes.
- Wear gloves and eye protection.
- Keep away from sources of ignition or direct heat.
- Ensure good ventilation and do not smoke when using this product.
- If splashed in eyes, wash away with plenty of water and seek medical advice immediately.
- Wash splashes off skin with plenty of water.
- Only sufficient 'Marclean GP' should be dispensed from original can to working containers for immediate work needs.

Graffiti removal procedure

1. Apply gel by suitable cloths to surface.
2. Work in by hand with soft bristle brush.
3. Leave for between 10 and 30 minutes, checking periodically for 'lifting' of graffiti.
4. Wash off surface with either hot or cold water from hose, or for large areas medium/high pressure water washer.
5. Repeat stages 1 to 4 if graffiti marks persist.

Limitations/precautions for surfaces

- Use only the specified removal agent, NOT gels containing chlorinated hydrocarbons as these will attack this paint system.
- It is possible there may be slight softening of the 2 pack paint after maximum exposure to the gel. The paint will recover its hardness within 2 hours.
- Do not use on conventional painted or plastic surfaces.
- Do not use abrasive grit or other abrasive products.

9 Graffiti removal in common target locations

Broadly speaking, target locations may be divided into the following categories:

1. building exteriors and interiors (depots, garages, warehouses, apartment blocks, sports and entertainment centres/arenas, council and other 'establishment' offices, schools, shops, shopping precincts, car parks, etc.);
2. open-air structures (bridges, walkways, fences, monuments and statues);
3. public lavatories;
4. train cars (exterior and interior);
5. buses and coaches (exterior and interior);
6. other buildings, surface and underground stations (exterior, interior, below-ground areas, platforms, trackside).

9.1 Building exteriors and interiors

Large public transport depots and garages are prime targets for graffitists since the graffiti crews know that trains or buses are stabled there, and that graffiti are removed from the trains and buses in those locations. The premises, therefore, seem to them to be bastions of the establishments opposed to their 'art' and thus a legitimate target. Many depots, garages, factories, warehouses and public buildings are particularly vulnerable to graffiti because their size and location, and often a shortage of manpower, make security and surveillance difficult, particularly at night when they are not fully occupied and when most graffiti crews strike.

Large developments such as shopping malls and sports centres very often attract graffitists who seem to make a bee-line for them when they are newly opened. Material specifications for the construction of large public developments should, wherever possible, anticipate the possibility of graffiti attack by calling for graffiti-resistant materials and coatings.

There is a case for not removing graffiti from the exteriors of buildings on industrial estates and from other structures that are not seen by many people. The high cost and low aesthetic value of removal could justify limiting, or even dispensing with, countermeasures. On the other hand, graffitists who get into bus and train depots and garages are sometimes able to get at the vehicles as well. Improved security systems in those places might, therefore, produce the double benefit of protecting both premises and vehicles.

Even small private shops (Plate 17) and infant schools (Plate 18) are not spared from attack, and in some areas, large blocks of flats suffer from graffiti inflicted by their own residents or people living nearby. The most vulnerable parts are those that cannot be seen from outside, such as stair wells and lift interiors. The visual impact of high density graffiti such as exists in some blocks of council flats can be very disturbing for both residents and visitors (Plate 19).

9.2 Open-air structures

9.2.1 Bridges, walkways and fences

A wide range of this sort of open-air structure is at risk from graffitists. Both masonry and metal bridges, walls and fences – especially those alongside railway lines, and isolated huts and other small buildings are favourite targets. Graffitists want their handiwork to be highly visible and access to the structures is easy, even if it is sometimes dangerous, as in the case of some bridgework and trackside constructions.

The main problems of graffiti removal and protection in such locations are:

- the size of many of the areas affected;
- safe access for graffiti removal or surface protection;
- lack of services (electricity, water, drainage) in remote sites;
- unsuitability of some methods near live track, over roads etc.;
- screening graffiti-free surfaces from repeated attack long enough for full cure of protective coatings or sealant to be achieved.

High pressure washers may be suitable for some open-air sites, such as bridges, provided sufficient water is available, possibly in tanks.

It may be feasible to partition off sections of the site while allowing public access to most of the area. Specialist advice on the best type of protective coating to adopt will probably be needed because of the wide range of construction materials used in open-air structures. Anti-corrosive barriers will of course be needed for steel structures.

As a result of being in the open air, these structures are well ventilated. Exposure to solvent vapours from protective coatings and graffiti removal agents will, therefore, be lower than in confined spaces. Nevertheless, compliance with health and safety regulations, suitable protective clothing and adequate supervision are essential.

Another favourite target for graffitists is advertising posters on billboards and hoardings, on walls inside public buildings and on the inside and outside of public transport vehicles. The simplest solution to this problem is usually to paste new posters over the defaced ones or, in the case of vehicles, prompt removal using an approved method.

9.2.2 Monuments and statues

Ancient monuments such as Stonehenge, early Tudor buildings and valued public statues are not immune from graffiti attacks. Whether the structure is in an isolated place or in a busy town, the surrounding area could still be congested by sightseers, so self-contained equipment is likely to be needed, hoardings may have to be erected and the work of graffiti removal may have to be done at a time when few people are expected to be present.

Where a monument is the subject of a preservation order, the original materials of construction will have to be preserved, and specialized advice may have to be sought before cleaning is started. If a protective coating is applied, it must be muted enough to preserve the appearance of the structure.

9.3 Public lavatories

Public lavatories present a unique problem where graffiti are concerned. Most of the graffiti are to be found inside the WC cubicles where, in seclusion, graffitists have leisure to indulge their creative whims without fear of being disturbed. Nevertheless, the walls outside WC closets in washrooms are also sometimes targets for graffiti (Plate 20).

Since most lavatory graffiti consist of obscene writings and drawings, they are offensive to many people who are hardly in a position to ignore them. The sheer volume of such graffiti also makes the problem expensive to deal with. Public lavatories are consequently obvious locations for graffiti resistant materials of construction and protective coatings.

9.4 Train cars

9.4.1 Car exteriors

Freight and passenger cars (commonly called 'carriages') often have bare metal exteriors. They are major targets for graffitists' 'tags' both in aerosol spray paint and bold felt-tip marks. Such graffiti, if fresh, can generally be completely removed with a removal gel followed by medium pressure (1500 psi) water washing (Plates 9–12). Non-gelled removal substances give inadequate contact time for effective removal.

Stubborn ghosting may remain if the graffiti is old or on surface-weathered cars. Further applications of gel and pressure washing may be effective. Removal of ghosting by mechanical or manual abrasive means is not recommended. It is very time consuming and it alters the surface texture of the metal.

Removal gel that is effective on bare metal may not be suitable on cars with painted exteriors as the gel may attack the paintwork. Painting over the marks with a surface coating that blends with the original paint is often the best solution. In London and elsewhere, new urban trains are now being painted with graffiti-resistant polyurethane paints and existing trains are similarly treated when they are refurbished.

9.4.2 Car interiors

Graffiti inside cars is particularly unpleasant. Passengers are faced with unsightly and sometimes obscene or racist scrawls. It may cause unease or fear, especially to women travelling alone or at night.

The problem of graffiti removal from car interiors is complex. A wide variety of material types may be affected. These include glass, melamine, painted panels, moquette and leather upholstery, stainless steel, textured plastic claddings and wood. Most of the graffiti is applied with felt-tip pens and aerosol paints on glass, melamine and painted panel surfaces, but there have been cases of indiscriminate aerosol spraying of car interiors, including upholstery.

Aerosol paint graffiti can normally be completely removed from glass by sharp metal-bladed scrapers and graffiti removal solvent. Mildly abrasive cream or suitable solvents will usually erase graffiti from untextured melamines. Textured melamines and other textured but non-permeable surfaces may be similarly treated, but the graffiti removal substance will require more working in with bristle brushes or swabs and will have to be thoroughly rinsed off afterwards. No electrical equipment and only small quantities of water are usually needed for such a straightforward manual operation which can be carried out reasonably quickly. Cars from which graffiti are to be removed should, of course, be taken out of service. The work should be done in the open air with doors and windows open for maximum ventilation when solvents are being used.

Coarse abrasive powder should never be used to remove graffiti from car interiors as it produces soiling and may adversely affect the surfaces being cleaned. Wire wool, or highly flammable solvents such as thinners should also never be used.

9.5 Buses and coaches

9.5.1 Bus exteriors

Graffiti attacks on the exterior of buses are quite common. Removal of the markings is a problem because of the damage solvent-based cleaners may do to the painted surfaces. The best available method at present is probably the cautious use of solvent-based graffiti removers with minimal contact time, followed by medium pressure water-jet washing. Matching colour paints may also be used to hide graffiti or restore surfaces after treatment. Good quality graffiti and solvent-resistant polyurethane lacquers in colour-matched tints are now available commercially but they must be applied by specialists.

9.5.2 Bus interiors

On double-decker buses, interior graffiti attacks are mostly made on the upper decks (Plate 3) or in the stairwells. As with trains, a wide range of material types and surface finishes may be affected by graffiti. Glass (windows), painted panels (ceilings, side pieces, some seat backs), plastic claddings and melamines are particularly prone to attack.

Graffiti removal should be carried out manually, while the bus is out of service, in the manner described above for car interiors.

Many plastic claddings are permeable to graffiti marks and it is almost impossible to clean them completely. Sometimes, bus companies have stripped the cladding from seat backs and kick boards, and painted over the exposed surfaces. This has permitted graffiti obliteration by over-painting, but graffiti marks will permeate some paints and may leave ghosting. Here again, and on ceilings and other areas, non-permeable polyurethane paints would give more protection.

It is very difficult to get rid of graffiti from bus seat upholstery, but the problem is likely to be eclipsed by more extreme acts of vandalism such as seat slashing and arson. A solution, devised by some authorities, is the replacement of upholstered seats most prone to the attentions of graffitists and vandals, by moulded plastic seats.

9.6 Transport buildings, surface and sub-surface stations

9.6.1 General considerations

Whenever possible, when selecting materials for the refurbishment or modernization of bus and train stations, airports, ferry terminals and certain other buildings, the potential graffiti problem should be taken into account.

Materials with high inherent graffiti resistance, such as ceramic tiles and vitreous enamel, require no additional surface protection treatment, but where materials such as terrazzo with poor graffiti resistance are to be used, it may be possible to apply protective coatings to pre-cast sections before installation in the station.

Often, however, existing interior finishes will have to be cleaned and protected *in situ*. For health and safety reasons, and also to prevent graffitists striking before the work is complete, areas will have to be partitioned off from the public. This requires careful planning and the consent of all interested parties to ensure:

- minimal disruption to passenger flow in stations;
- maintenance of the safety of walkways, exits, lifts, escalators and platforms for the public and staff, particularly at rush hours and in the event of emergencies;

- adequate ventilation, keeping toxic and flammable vapours within acceptable limits, both for the public and for employees working on graffiti removal or surface protection;
- keeping noise and vibration to acceptable levels;
- adequate water or solvent waste removal facilities;
- safe storage for materials and equipment;
- the provision of necessary services (water, drainage and electrical supply of the correct type);
- good fire safety for the anti-graffiti protective coating.

With careful planning, it should be possible to work on a location, section by section, using movable barriers and screens, and ensure that surfaces are re-exposed only when they are fully protected from permanent graffiti marking. Sections to be cleaned or protected should be kept small enough to restrict the volume of water, solvents, cleaners and lacquers to a minimum, consistent with efficient working.

In many cases, where the area affected is small or the problem is not too severe, it is possible to use manual methods to remove graffiti and protect the surface. This eliminates the need for electrical machinery and considerably reduces the problems of noise, dust, fumes and the supply of electricity. People engaged in the work will probably have to carry it out during overtime or at night at busy locations.

9.6.2 Exteriors

Methods of dealing with graffiti on building exteriors depend on the building's situation and the materials used in construction. Buildings in heavily built-up areas pose different problems, in terms of work access and protective barriers, from those that are detached from other properties or are remote from traffic or public access.

Properties in congested areas are less prone to graffiti attack than isolated ones because access is usually more difficult and the possibility of being seen is greater. There are, however, exceptions.

Since exterior building materials are exposed to the atmosphere, as well as to the attentions of graffitists, it is important to ensure that cleaning and protective work does not diminish either the weather protective function of the surface or the aesthetic integrity of the building. Often the areas affected are large, and rapid mechanical methods of removal and cleaning may be the most effective. But excessively harsh treatments, such as grit-blasting, must be avoided as these could deface the building. Techniques not permissible elsewhere may, however, be possible in these exterior locations because they are well ventilated and offer facilities such as good waste water drainage.

If an anti-graffiti protective coating is considered necessary, it is obviously best to apply it before an attack is made. Failing that, surfaces must be clean and dry before applying a coating, and the area will need to be protected by screens from both weather and further graffiti attack until the coating is fully hardened.

9.6.3 Interiors

Interior locations at risk include such areas as concourses, subways, escalators, staircases and lifts, station ticket halls and waiting rooms.

In locations that are large in relation to the flow of people, it should be possible to partition off areas sequentially without too much inconvenience to foot traffic. Care must be taken to ensure that the public and staff are not at risk from toxic fumes, dust or spilled water, nor subjected to too much noise. In addition, partitions must be kept in good order to protect the public and prevent further graffiti attacks.

It is desirable to finish graffiti removal and protection work as quickly as possible in those areas frequented by the public, so overtime or night work may be required.

Many different material surface types are to be found inside public access buildings. They include emulsioned render, terrazzo, travertine marble, ceramic tile, painted and unpainted woodwork, glass and decorative plastics. Although there may be a lower incidence of graffiti attack than in some other locations because of the greater chance of graffitists being seen by the public and staff, the range of removal problems may be wider.

9.6.4 Below-ground areas

In many below-ground locations, the use of some graffiti removal techniques is not feasible because of access problems for heavy equipment and lack of ventilation and drainage. Most of the work has to be carried out at night or at other times when the station is closed. If anti-graffiti coatings are applied below ground, both the work and the materials used must conform to fire safety regulations.

Plate 1 Graffitist psychology — graffiti denounced as 'toy'.

Plate 2 Vandalism and graffiti at a bus station. Smooth painted surfaces are a target for felt tip markers.

Plate 3 Graffiti on a bus interior — most surfaces are attacked.

Plate 4 Graffitist psychology — a 'hall of fame' for tags on a residential building exterior.

Plate 5 Fly posters in a public walkway. The smooth mosaic surface accepts both posters and felt tip markers.

Plate 6 Typical contents of a portable graffiti-removal kit. The contents include: *Back row* — selection of appropriate graffiti removal products; *Middle* — cleaning tools (scrapers, swabs, brushes); *Foreground* — vapour masks, protective gloves, goggles, used swab disposal sack.

Plate 7 Mechanical work-in of graffiti removal agent with electric rotary brush may speed up its action.

Plate 8 Mobile medium pressure water washer and lance with diesel powered water heater.

Plate 9.

Plate 10.

Plate 11.

Plate 12.

Plate 9 Graffiti on exterior of aluminium train.

Plate 10 Application of gelled graffiti remover to aluminium train exterior.

Plate 11 Washing off graffiti and removal agent with a pressure washer.

Plate 12 Aluminium train after graffiti removal.

Plate 13.

Plate 14.

Plate 15.

Plate 16.

Plate 13 Heavy graffiti attack with felt tip pen and aerosol paint on unprotected terrazzo. Note the ghosting from previous removal attempts.

Plate 14 Terrazzo in Plate 13 partially cleaned of graffiti. Only stubborn marks remain.

Plate 15 Re-application to (left) and removal (right) of graffiti removal mortar from the terrazzo in Plate 13.

Plate 16 Terrazzo in Plate 13 cleaned of graffiti and protected with a clear satin finish barrier coating.

Plate 17 Shop front hit by graffiti. Phone booth used as a work platform for 'piece'.

Plate 18 Graffiti on young children's playschool, mainly aerosol paint on heavily textured 'pebbledash' walls.

Plate 19 A focal point for graffiti — a staircase in a high rise apartment block.

Plate 20 Graffiti in toilets — less obvious aspects. The unprotected wood of doors exteriors selected as a target.

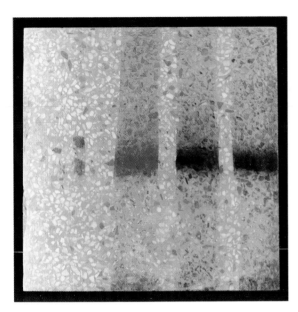

Plate 21 Poor graffiti resistance of unprotected terrazzo — the surface is smooth but permeable.

Plate 22 Good graffiti resistance of semi-glazed brick. The surface is textured but not porous, and the top halves of the stripes have been easily removed.

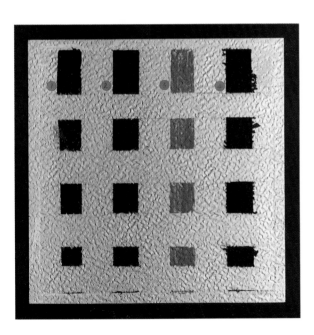

Plate 23 Good graffiti resistance of textured but unweathered aluminium. Gaps in the vertical stripes are where the graffiti have been easily removed.

Plate 24 Appearance of 'permanent' anti-graffiti protective coatings on porous brick: (a) uncoated; (b) permanent coating A (satin); (c) permanent coating B (gloss); (d) permanent coating C (satin).

Plate 25 Appearance of 'sacrificial' anti-graffiti protective coatings on porous brick: (a) uncoated; (b) sacrificial coating A; (c) sacrificial coating B.

Plate 26 Anti-graffiti protective coatings on porous brick — layout of test graffiti marks of aerosol paint and felt tip.

Plate 27 Appearance of porous bricks protected with sacrifical coating B after removal of graffiti (see Plate 25).

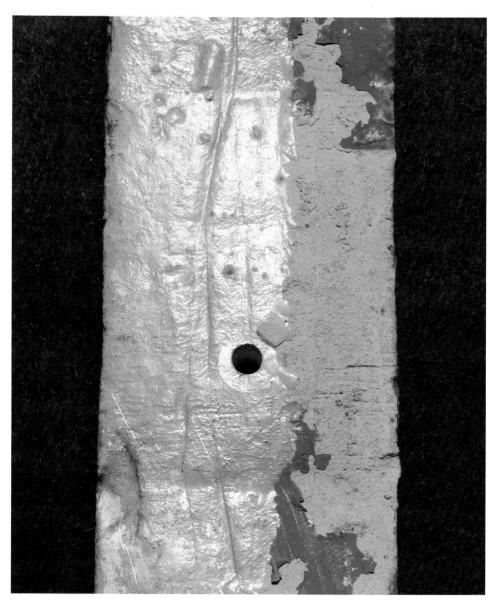

Plate 28 Porous brick protected with 'permanent' coating. The coating has been destroyed after one graffiti removal cycle.

Plate 29 Painted wall and door attacked by graffiti after protection with sacrificial coating B.

Plate 30 Removing graffiti from protected surfaces in Plate 29 using hot pressure wash only.

Plate 31 Graffiti completely removed, but protected surfaces need replenishment of sacrificial coating.

Plate 32 Application of anti-graffiti protective coating to terrazzo by roller and brush.

Plate 33 Interior of self-contained graffiti removal van showing equipment and removal agents. (Photograph courtesy of Graffiti Clean.)

Plate 34 Poor graffiti removal from painted surfaces. Inappropriate solvent product has been ineffective and has increased the damage.

Plate 35 Poor graffiti removal from porous brickwork — use of the wrong method leads to only partial removal of aerosol paint.

Plate 36 Obliterative overpainting often encourages further graffiti attack and may look unsightly.

Plate 37 Painted murals may not deter graffiti.

Plate 38 Suitable plants may hide graffiti or make access to walls difficult.

9.6.5 Platforms and trackside

Clearly, all formal procedures and regulations relating to work on platforms and trackside apply when graffiti removal and protection operations are in progress close to where trains run. If there is any possibility of rail traffic running, look-out personnel will have to be posted, and possession orders will have to be obtained, particularly for electrified track to isolate the track section being washed on.

When graffiti removal or surface protection chemicals are applied to electrical or signalling equipment, care must be taken to ensure that the equipment retains its essential properties in terms of insulation or conductivity. Expert advice will be required before starting such work.

10 Surface type and graffiti removal

When graffitists choose their surfaces and marking materials, their purpose is the direct opposite of that of the owners of the targets – to ensure that the graffiti are as permanent as possible.

Both the type of material making up a surface and its texture affect not only the prominence and durability of the markings but their ease of removal. It is, therefore, essential when planning a graffiti removal or protection operation, to pay attention to the type and texture of the surface material.

When new buildings or public transport vehicles are being designed and built, and when premises are to be extensively refurbished, materials and surfaces should be specified with graffiti removability in mind. If this cannot be done because of considerations of construction or appearance, surface protection treatments may be needed.

10.1 Types of surface material

The most important properties of surface materials where ease of graffiti removal and durability of finish are concerned are as follows.

1. *Non-permeability*. The material's micro-structure does not allow the dyestuffs and pigments from graffiti-marking devices to penetrate the surface.
2. *Hardness*. The material resists scratching or abrasion of the surface.
3. *Resistance to chemical cleaning agents*. The material withstands repeated cleaning with solvents, detergents and other chemical agents without adverse effect to the surface.

4. *Resistance to weathering.* The material stands up to a wide range of temperatures and, in outside locations, prolonged exposure to rain and sunlight.

Among materials that meet those criteria are glass, glazed ceramic tiles, vitreous enamel and stainless steel. They are generally smooth in texture, although some sectional profile and embossing may be deliberately introduced during manufacture, as in reeded glass.

In assessing the inherent graffiti resistance of a material, the surface texture must always be taken into account. This is discussed in more detail in section 10.6, 'Surface texture and graffiti resistance'.

10.2 Permeable surface materials

Materials with high permeability to dyes and pigments usually have poor graffiti resistance, even when the surface is smooth and apparently non-porous. Once the dye has penetrated the material, it can be very difficult to remove (Plate 21). Many commonly used materials are inherently permeable. Among these are PVC, and gloss, emulsion and other paints.

Similarly, porous materials, such as terrazzo, concrete, grouting, mortar, granite and common brick stock, that have been subjected to graffiti are also particularly difficult to clean.

Because it is not always possible to determine the precise material type by non-scientific means, it may be useful to refer to the original building specifications or to make laboratory tests before attempting to take remedial action. A surface of conventional alkyd-based paint is visually indistinguishable from superior colour-matched polyurethane of certain sheens, but they will react very differently to graffiti and cleaning chemicals.

To counteract the threat of graffiti, it may be possible to replace proposed or existing permeable materials with non-permeable surface cladding. If that is not possible, then non-permeable barrier or sealant protective coatings could alleviate the problem. However, protective coatings must be chosen with extreme care. Many so-called sealants on the market fail to provide adequate anti-graffiti protection, either because of their own inherent permeability and poor graffiti resistance or because of their poor durability which allows subsequent re-exposure of vulnerable surfaces.

Indeed, permeable plastic-based materials, such as vinyls and paints, often have poor hardness – and therefore poor durability – as a result of their open molecular structure. Repeated cleaning with solvent or abrasive-based agents may also cause them to lose plasticizers of dyestuffs built into the original formulation. The leaching out frequently leads to deterioration of the natural colour and texture of the treated surface. This is invariably aesthetically unacceptable and will make the surface even more susceptible to staining and ghosting.

10.3 Hardness of surface materials

Materials that are not hard or coherent enough may suffer from cutting or scratching. They may also be damaged by abrasive products used in attempts to remove graffiti. Even if the outer skin of the surface material is impermeable, breaking the barrier through scratching, cutting or abrasion may expose permeable or porous underlying regions which irreversibly take up graffiti-mark colourants, and so make complete removal impossible.

If the surface is friable, its durability will be poor and it is likely to be permeable or porous to graffiti-marking colourants, posing the same problems of removal. Harsh mechanical methods, such as linishing or shot blasting, applied to crumbly or soft materials are likely to leave unsightly scars or, at best, achieve a finish that is aesthetically disagreeable.

Ageing or weathering may turn the exteriors of hard materials into crumbling or porous surfaces. This often happens to old buildings and even occurs as a microscopic layer on unprotected aluminium exteriors. In both cases, the most widely accepted method of graffiti removal and general cleaning of reasonably large areas is by chemical agents and medium pressure water jets.

10.4 Surface resistance to chemical cleaners

Permeable or crumbly materials are likely to have poor resistance to certain chemical agents because they soak them up and retain them within the surface for a long time. This is particularly true of the weathered exteriors of stone buildings. Deterioration can be retarded by removing the absorbent deposits on the surface with an appropriate chemical cleaning agent followed by low to moderate pressure water rinsing with fan jets. This should also remove the bulk of any graffiti on the crumbly surface layer.

Use of incorrect chemical agents and water rinsing methods may, however, greatly increase the damage. Expert advice should be sought, especially for the exteriors of buildings of special architectural importance (e.g. listed buildings).

It is essential to choose the right cleaning solvents for plastic or painted surfaces. Many solvents that dissolve felt-tip and aerosol marks also have a deleterious effect on conventional paints and many plastic surfaces by softening, leaching out colour or altering surface finish.

10.5 Surface resistance to weathering

Material surfaces with good resistance to weathering are usually not permeable or porous, and they normally have adequate hardness and good resistance to chemicals in the atmosphere.

Conversely, porous masonry materials, such as stone cladding and common brick, may be rapidly weathered by aggressive chemicals in urban rainwater, or through spalling by absorbing water and freezing in winter. Clearly, any material selected for graffiti resistance in outdoor locations must have good resistance to weathering if graffiti protection is to be durable.

With some exterior locations, it is possible to reduce porosity and improve surface hardness with appropriate surface coatings. Such coatings, however, must not interfere with the natural behaviour (permeation and thermal movement) or the underlying substrate. Also, where large areas are involved, the cost of such coatings may be prohibitive in relation to the aesthetic impact.

10.6 Surface texture and graffiti resistance

The surface texture of any material type may significantly affect ease of graffiti removal and general cleaning of a surface. The texture must be taken into account when assessing a graffiti removal or surface protection task.

Texture is notoriously difficult to describe unambiguously and often more difficult to measure quantitatively. In an attempt to resolve this, terms such as surface texture, surface profile and surface irregularities and defects are explained and discussed below.

Irregular textures (natural)

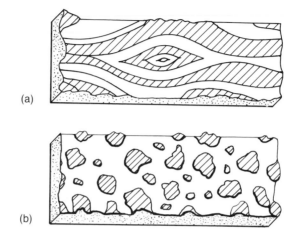

(a)

(b)

Regular textures (embossed patterns)

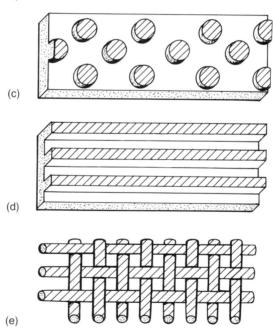

(c)

(d)

(e)

Fig. 10.1 Surface textures: (a) woodgrain; (b) concrete; (c) studded; (d) ribbed; (e) fabric/woven. (*Note* Shaded areas imply surface above plane of paper.)

10.6.1 Surface texture

Surface texture is, in general terms, the overall physical state of a surface assessed visually or by touch. It may be described as matt, highly polished, rough, smooth, ribbed and so on.

The final texture of a material may be inherent in its natural state, e.g. wood grain and crystalline rock. It may be imparted by processing such as grinding or polishing, or it may be imprinted on synthetic materials during manufacture, as in wood grain melamines and leather-finish vinyl sheet. Regular, patterned or irregular textures may be achieved by such imprinting (Fig. 10.1).

10.6.2 Surface profile

The surface profile relates to the cross-sectional topography of a surface. This may be visible either with the naked eye or under the microscope. It is important because it may strongly influence how the surface will respond to cleaning.

Profiles may be regular in machine-embossed patterned surfaces, or irregular, i.e. natural (Figs 10.2 and 10.3). Important aspects of the profile are listed below:

1. the relative height of peaks and depth of valleys in the surface – shallow profiles are easier to clean;
2. the spacing of peaks relative to valleys – very open texture patterns allow easier access of cleaning agents and tools, such as brushes;
3. the general shape of the profile – gentle smooth peaks and valleys also allow easier access by cleaning agents and tools than do ragged ones that may trap paint, pigment, and so on, making removal difficult.

The above criteria apply to surface profiles whether they are clearly visible or can be seen only under optical magnification, for example weathered stone or metal surfaces. The eye or hand usually discerns microscopic textures as matt or 'roughish', respectively.

Regular or even surface textures may be amenable to semi-quantitative measurement, using a surface roughness analyser such as the 'Surtronic 10'. This is a small hand-held instrument with a stylus which will traverse the surface and measure its undulation. For good results, a number of measurements must be taken. The instrument does not work on curved or very uneven surfaces.

Table 10.1 gives some typical values of roughness (R_a) for a range of surface types.

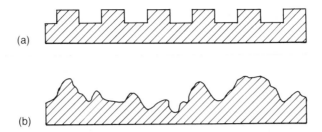

(a)

(b)

Fig. 10.2 Diagrammatic surface profiles: (a) regular; (b) irregular.

Regular

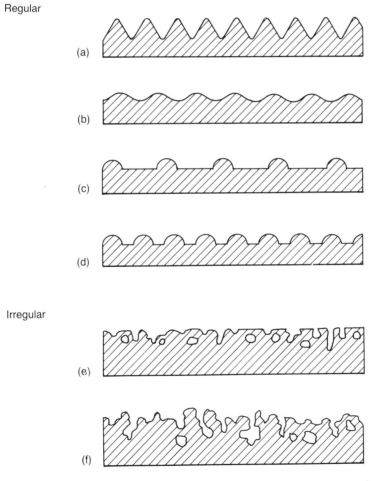

(a)

(b)

(c)

(d)

Irregular

(e)

(f)

Fig. 10.3 Diagrammatic aspects of surface profile: (a) deep; (b) shallow; (c) open; (d) closed; (e) porous but smooth; (f) porous and rough.

Table 10.1 Some typical roughness (R_a) values for materials surfaces

Material	Surface finish	R_a (μm)*	Notes
Stainless steel	Highly polished	0	General usage
White high glaze ceramic wall tile	Smooth	0.26	
Plain glass	Polished	0	
Aluminium	Weathered (semi-matt)	2.2	
Aluminium	Linished (100 grit)	1.8	
Aluminium	Glass bead peened	3.2	
Melamine	Smooth (new)	0.3	
Melamine	Unprotected/abraded (old)	0.7	Interior wall claddings
Melamine	'Matt' finish	0.3	
Melamine	'Suede' finish	1.3	
Melamine	'Quarry' finish	3.3	
Safety floor	Gritted finish	4.7	Floorings†
Cork flooring	Rough, irregular	8–9.5	
Linoleum	Matt	2.7–2.9	

* Measured on 'Sartronic 10' available from Rank Taylor Hobson Ltd, PO Box 36, New Star Road, Leicester LE4 7JQ, UK. Tel.: (0533) 763771.
† Floorings usually require significant 'roughness' or texture to retain good slip-resistant properties.

10.6.3 Surface irregularities and defects

Surface irregularities such as cavities, pits, voids, cracks, crazing and blisters may be naturally present in surface materials, or they may be there as a result of manufacturing faults. They could also occur as wear and tear on the surface of man-made materials. Such defects may eventually lead to deterioration of the surface to a point where cleaning becomes very difficult.

10.6.4 Smooth surfaces

Smoothness is relative. A surface that seems smooth to the touch or because it has a high gloss is not necessarily smooth microscopically, nor is it bound to be non-porous or impermeable. Nevertheless, materials that are essentially hard, non-porous and impermeable are much more easily cleaned if they are smooth.

Instruments such as the Surtronic 'Roughness' meter make, within distinct limits, semi-quantitative measurements of smoothness. Such instruments are useful for assessing slight deviations in the smoothness of a surface caused by mechanical working or abrasion. In some cases, the measurements can be related to cleanability. Two scales of measurement are used: R_a (units in μm) and R_{tm} (RzDIN) but the results must be interpreted with great caution. On the R_a scale, highly polished stainless steel gives an ostensible reading of 0.0, high glazed 'smooth' ceramic tiles 0.0–0.3 and weathered aluminium in the region of 2.0–2.2.

Highly textured and very rough or irregular surfaces are not amenable to measurement by such instruments.

11 Graffiti resistant surface materials

Appraisal of the material type and texture of exposed surfaces makes it possible to assess or predict the inherent graffiti resistance, and amenability of graffiti removal, of many materials commonly used in the construction of buildings, other structures and vehicles.

11.1 Inherent graffiti resistance of material surfaces

Table 11.1, which is admittedly neither exhaustive nor definitive, summarizes the results of some materials appraisals. In most cases, graffiti resistance has been assessed through laboratory trials or on-site experience. As a general rule, hard, non-porous and impermeable surfaces are likely to have very good inherent graffiti resistance and cleanability (Plates 22 and 23). Permeable or porous surfaces may not easily allow graffiti removal, and rough or heavily-embossed textures prevent thorough cleaning of the surface.

For materials where inherent graffiti resistance is very good, application of anti-graffiti protective coatings is unlikely to be beneficial, and indeed may be detrimental. Conversely, materials with poor inherent graffiti resistance (e.g. brick, cement render) are best protected before graffiti attack, and this is increasingly being specified for newly-constructed buildings.

Graffiti resistant surface materials

Table 11.1 Inherent resistance of material surfaces

Material category	Material quality	Surface type	Ease of removal	
			Permanent marker	Aerosol paint
Metal				
Aluminium (weathered, exterior)	Medium-hard impermeable	Semi-smooth	Fairly good	Fairly good
Stainless steel	Hard, impermeable	Smooth to very smooth	Good	Good
Vitreous				
Glass (clear)	Hard, impermeable	Very smooth	Very good	Very good
Ceramic tile (hard glaze)	Hard, impermeable	Semi-smooth to very smooth	Very good	Very good
Vitreous enamel	Hard, impermeable	Smooth to very smooth	Very good	Very good
Stove enamel	Fairly hard, impermeable	Smooth	Good	Good
Masonry				
Concrete	Medium-hard, porous	Semi-smooth to rough	Poor	Fair
Brick: (common stock)	Medium-soft, porous	Rough, crumbly	Poor to very poor	Fair to poor
Brick: semi-engineering (glazed)	Medium-hard	Semi-smooth to rough	Fairly good	Fairly good
Terrazzo	Medium-soft, porous	Semi-smooth	Poor	Fair
York stone	Medium-soft, porous	Rough	Poor	Poor
Travertine Marble (unfilled, unsealed)	Soft, very porous	Many cavities	Very poor	Very poor
Marble mosaic	Medium-soft, porous	Semi-smooth (many joints)	Fairly poor	Fairly poor
Pointing (cementitious)	Soft, porous	Rough, crumbly	Very poor	Very poor
Grouting (cementitious)	Soft, porous	Semi-smooth, crumbly	Very poor	Very poor
Grouting (epoxy)	Medium-hard,	Semi-smooth	Good	Good

Table 11.1 *Continued*

Material category	Material quality	Surface type	Ease of removal	
			Permanent marker	Aerosol paint
Plastics				
Glazing 'Perspex'	Fairly soft, impermeable	Smooth	Fairly good	Fairly good
Polycarbonate	Fairly soft, permeable	Smooth	Poor	Poor
PVC (clear)	Fairly soft, permeable	Smooth	Poor	Poor
Laminates				
Melamines	Hard, impermeable	Semi-smooth to smooth	Good	Good
Glass-reinforced plastics	Fairly hard, impermeable	Semi-smooth to smooth	Good	Good
Vinyl claddings	Soft, permeable	Smooth to heavily-textured	Very poor	Very poor
Flooring				
Rubber	Soft, permeable	Smooth to heavily textured	Poor	Fairly good
Linoleum	Fairly soft, permeable	Smooth to heavily textured	Fairly poor	Fairly poor
Paints				
Gloss	Soft, permeable	Smooth	Very poor	Very poor
Emulsion	Soft, permeable	Semi-smooth	Very poor	Very poor
Exterior (weatherproof)	Fairly soft, permeable	Rough	Poor	Poor
Powder coatings	Fairly hard, semi-permeable	Smooth	Good	Good
Epoxy paints	Hard, impermeable	Semi-smooth to smooth	Fair to good	Fair to good
Polyurethane lacquers	Hard, impermeable	Smooth	Very good	Very good

* 'Perspex' is a trade name for polymethyl methacrylate.

95

12 Anti-graffiti protective coatings

Many different types of coating and system are available, ranging from a single coat of clear lacquer to a three or four-coat pigmented system. Nearly any surface may be coated to improve its graffiti resistance, and a wide variety of decorative finishes may be achieved – solid colour, multicoloured or textured.

In all cases, the degree of graffiti resistance depends upon the nature of the top coat. The various systems available are, therefore, often categorized by the binder contained in the top coat.

12.1 Clear (unpigmented) coatings

Clear coatings are used where it is important to maintain the natural appearance of the substrate. They are available in a range of gloss levels to suit the original finish (Plate 24).

The most durable types available in this category are moisture-curing and two-pack polyurethanes. These will normally withstand many applications and erasures of graffiti and suffer no adverse effects. There are also a number of water-borne acrylic lacquers on the market, but their graffiti resistance is limited, and the graffiti removal agents are liable to soften them severely.

12.2 Pigmented coatings

Pigmented, or coloured systems perform the dual function of obliterating existing graffiti and providing a surface from which graffiti may be readily removed. They are usually in the form of multi-coat systems comprising primer, undercoat and finish, although there are a number of variants.

There are several different types of coating in this category, including moisture-curing and two-pack polyurethanes and water-borne and solvent-borne epoxies. As with the clear coatings, the pigmented polyurethanes are the most durable of the materials available.

Solvent-borne epoxies may have similar graffiti resistance to polyurethanes, but they have limited stability in ultraviolet light. Water-borne epoxies, unlike those that are solvent-borne, present no vapour hazards but are less durable.

12.3 Sacrificial coatings

There is an increasing interest in sacrificial (temporary) coatings. These coatings are designed to be easily, or partly, removed together with any graffiti they carry, and then later replenished.

At first sight, it might seem that permanent anti-graffiti coatings are preferable to sacrificial ones, on the grounds that they last longer and that labour costs of application are therefore lower. However, it must be remembered that some so-called 'permanent' coatings have limited durability (Plate 25). Even polyurethane coatings may get damaged after graffiti have been cleaned off them a number of times, particularly if wrong cleaning methods have been used, and removal and replacement of damaged polyurethane may be difficult and costly.

A cost-saving advantage of sacrificial coatings is that often only pressure washing with water is needed to remove graffiti from the surfaces they cover (Plates 26 and 27). The use of expensive chemical solvent graffiti removal agents is thus avoided. In addition, there are fewer health and safety problems, and the costs of measures to protect people from possible contact with solvents are eliminated.

Also sacrificial coatings need not necessarily have a short life. Usually, only a part of the protective coat is stripped during graffiti removal by pressure washing. This can be restored by touching up the coat. Some systems are guaranteed by their manufacturers to last five years. Sacrificial coatings could be a better economic proposition than conventional ones, provided that recommended maintenance procedures are followed, preferably by contractors who both remove graffiti and apply the coatings (Plates 29–31).

There are two main types of sacrificial coating:

1. solvent-based silicone/wax coatings;
2. water-based polysaccharide coatings.

Both types of coating have a good appearance compared with other coatings. Water-based ones in particular, being essentially invisible, do not interfere with the natural look of the substrate (Plate 25).

Silicone/wax coatings acquire a sheen if they are rubbed repeatedly, so it is prudent to prevent contact by passers-by. This can be done by placing objects such as litter bins, seats and advertising frames in front of walls coated with this system. If this tendency of silicone/wax coatings to become polished when they are rubbed is a disadvantage, there are two compensating benefits:

1. because of their waxy nature, it is difficult to write on them with felt-tip pens, so they discourage some graffitists.
2. water-based adhesives cannot easily wet a wax coating. This means that normal fly posters will not adhere well to the surface.

Water-based sacrificial coatings, unlike silicone/wax types, can be reapplied to touch up the surface immediately after graffiti removal while the surface is still damp.

A further advantage of sacrificial coatings is that, unlike polyurethane, they are permeable and permit the release of moisture held in the building materials they cover. This may prevent premature weathering and spalling of stone or brickwork.

12.4 Obliterative coatings

Obliteration, either as a permanent solution or as a stopgap until longer-term action is taken, can also be considered as an anti-graffiti option. No materials are manufactured specifically for such purposes, but many substances exist that may be used to cover graffiti. The choice of material will depend on several considerations including speed of application, effectiveness, cost and compatibility with materials intended to be applied later.

12.5 Textured coatings

Texture is normally achieved by incorporating aggregate in the coating or by using spatter or hopper guns to apply more viscous materials. Graffiti resistance is usually obtained by the final application of a clear lacquer.

In theory, heavily-textured surfaces discourage fibre tip and ball-point pen graffiti, but such surfaces are vulnerable to aerosol paints which are then difficult to remove.

12.6 Fleck coating systems

Fleck systems typically comprise solid colour basecoats in which contrasting colour flecks have been incorporated. They are usually spray-applied, although systems that may be brushed on are now available.

The graffiti resisting glaze applied as a finish coat is usually water-based and of very limited durability.

12.7 Flake coating systems

These are usually two-pack polyurethane systems comprising coloured basecoats onto which PVA flakes of a contrasting colour are applied. A clear two-pack finish gives a very durable graffiti-resistant surface. The flake appearance may also help camouflage some light graffiti.

13 Applying protective coatings

13.1 Surface preparation

Thorough surface preparation is vital for good adhesion of the protective coating to the substrate. Poor adhesion can result in the following:

- failure or reduced durability of the coating;
- delamination in the event of a fire, allowing the flame to spread rapidly across the coated surfaces.

13.1.1 General cleaning

All surfaces need some form of cleaning to remove dust, dirt, grease or other contaminants before coating. The cleaning method will depend on the type and extent of soiling and the nature of the surface.

Porous surfaces, such as render, brick and stone, usually require only a brush-down with a stiff bristle brush, but heavily-contaminated surfaces may need high pressure water or steam cleaning. Non-porous, or less porous surfaces – including most types of paint, glazed brick and tiles – may be cleaned by hand with detergent solutions, emulsifying agents or solvents, or by pressure washers incorporating various chemical cleaners.

It is important to remove any previously-applied wax, acrylic or other sealant completely, and all traces of cleaning agents, detergents and debris must be removed by rinsing with clean water and allowing to dry before applying the coating.

13.1.2 Abrading

Smooth, hard, non-porous surfaces, such as old oil-based paints, need thorough abrasion to promote good adhesion of the protective coating. This may be done by hand or with a power sanding device, using a suitable grade of abrasive paper.

13.1.3 Grit blasting

Grit blasting is used mainly for cleaning badly-corroded iron and steelwork. The effectiveness of the method depends on the air pressure and the type of abrasive medium. Grit blasting may also be used on other metallic and non-metallic surfaces, provided care is taken to avoid excessive damage to the surface.

13.2 Equipment and application methods

Factors influencing the choice of equipment and application method include:

- the type of coating;
- the surface profile;
- the required standard of finish;
- health and safety (particularly in respect of vapour release when solvent-borne coatings are applied).

13.2.1 Brush

Brushing is the simplest means of application. One advantage is that the rate of solvent release during the application of the coating is relatively slow. However, some graffiti-resistant coatings are unsuitable for brush application and others may give a poor finish through poor flow.

Equipment Good quality varnish brushes should be used for most coatings, but cheaper brushes or sash tools may be adequate for temporary or obliterative coatings.

13.2.2 Roller

Coatings can be applied faster with rollers than with brushes, and the rate of solvent release is still low. Rollers are also more effective than brushes in applying materials with poor flow characteristics.

The standard of finish achieved by certain materials, such as some moisture-curing polyurethane lacquers, may be improved if they are applied by roller and 'laid off' using a brush (Plate 32).

Equipment Many types of roller are obtainable. Suppliers should be able to advise on which type to use for particular coatings. Where the coating contains aggressive solvents, phenolic resin-impregnated sleeves may be used to protect the main roller sleeve. Disposable sleeves are usually cost-effective when applying solvent-based substances.

13.2.3 Spray

Most types of coating are amenable to spray application. The method is much quicker than brush or roller application on large flat surfaces, but the amount of masking needed may be too expensive for other smaller areas. In addition, the rate of solvent vapour release is much higher than when using brushes or rollers. This would preclude the use of spray coating with solvent-based materials in areas where ventilation is restricted.

Spray application gives a good finish in the hands of skilled operators.

Equipment A number of types of spray equipment are available for different coating materials. The coating manufacturer should be able to recommend suitable equipment, thinners, methods of application and general work practices.

13.2.4 Suppliers of coatings

Names, addresses and telephone numbers of some suppliers of graffiti-resistant coatings are given in Appendix A.4.

14 Contractual aspects

14.1 Comparative costs for graffiti removal and surface protection

Little accurate information exists on the true costs of removal of graffiti and protection of surfaces. Authorities that employ their own labour to carry out the work, and those that use contractors on a regular basis, should be able to say how much they have spent, but when cost/effectiveness is taken into account, it is not easy to identify the most economic course of action.

In assessing the costs for graffiti removal and surface protection there will be a number of variables that make the assessment difficult. For example, removal agents and protective coatings vary both in cost and efficacy; labour costs vary depending on the amount of supervision operatives receive and whether or not skilled people are used; contractors specializing, or claiming to specialize, in graffiti removal and the application of anti-graffiti coatings, differ considerably in their effectiveness; every authority and every contractor has its own individual way of working, its own range of equipment, its own choice of removal agents, paints and coatings and its own degree of commitment to the training of operatives.

Despite these obstacles to establishing the costs of graffiti removal and anti-graffiti protection, the information in Table 14.1 which details charges for three different ways of dealing with graffiti, gives an indication of comparative costs.

14.2 Choosing a contractor

An efficient and reliable contractor, with well-trained staff and good equipment, can do a better and less expensive job of graffiti removal and/or applying protective coatings than some authorities

Table 14.1 Guide to the comparative costs of the application of permanent or sacrificial anti-graffiti coatings, and the removal of subsequent markings. (Work carried out by approved contractors on rough-faced brick wall, 100 m^2.)*

	Case 1 *Clear polyurethane coating*		Case 2 *Solvent-based sacrificial coating*	
Application labour and materials	£8.50/m^2	£850	£9.00/m^2 (1 primer and 2 coats)	£900
Erecting and striking hoarding to protect surface during curing of coating†		£700		
Application – total		£1550		£900
Graffiti removal (10 m^2) Labour	(2 hours travel plus 4 hours on site) @ £15/hour	£90	(2 hours travel plus 2 hours on site) @ £15/hour	£60
Materials	5 litres chemical remover	£40	2 litres sacrificial coating (for re-protection)	£20
Removal – total		£130		£80
Total cost for 12 months (1 operation per month)	Initial protection 12 removals (12 × £130)	£1550 £1560	Initial protection 12 removals (12 × £80)	£900 £960
Total		**£3110**		**£1860**

Case 3 Cost for graffiti removal from an unprotected wall§ – £25/m^3.   10 m^2 × 12 visits **Total = £3000.**

* Costs as at October 1989.
† Case 1 only.
§ Some permanent damage inevitable.

using their own resources. Unfortunately, since some contractors are neither efficient nor reliable, it is important to carry out some checks before engaging a new contractor.

■ Check that the contractor has the necessary knowledge and experience. Ask for the names and addresses of previous clients, phone them and ask whether they were satisfied with the quality of the work done.

■ Examine the contents of the contractor's van (Plate 33). Does it have self-contained facilities such as a pressure washer in good condition and its own electric generator? (Some carry their own water supplies in a floor tank. Others arrange with local authorities to use hydrants.)

■ Ask to see material data sheets to check that acceptable graffiti removal agents, paints and so on, are to be used.

■ Check the owner's awareness of health and safety regulations.

■ If sacrificial coatings are in use or wanted, employ a contractor who both removes markings and applies the coatings.

■ Check that the contractor has adequately trained staff for the job. Some contractors may be able to produce recognized certificates of training or membership of trade bodies.

■ If you want a call-out service, check that it is offered. Is there a back-up service to maintain the condition of cleaned surfaces? What are the costs and response time?

■ Find out how the contractor proposes to protect passers-by, if relevant.

■ What are the possibilities of night-work and weekend working where only limited access time is available? Ask for a sample of graffiti removal/surface protection on the site under consideration.

After the work has been done, you will, of course, check its quality. It is sometimes useful to take photographs to support claims and for future reference. The photographs should be marked with both location and date.

Finally, remember that cheap work is seldom good work.

14.3 Faulty approaches to graffiti removal and surface protection

Graffiti may be eyesores, but just as ugly are the results of attempts to remove them by inappropriate or slipshod methods. Unfortunately, many efforts to get rid of the markings leave the surfaces looking quite as bad as they were before the work began. Often this is because the people trying to clean them have no idea of what agents and methods are required, or because they have not been properly instructed, or have ignored the instructions.

A faulty approach to graffiti removal may result in damage to the surface in any of the following ways.

■ Gloss or emulsion paintwork, PVC and some other plastics may be attacked by the potent solvents present in many graffiti removal agents, leading to softening, blistering or destruction of the surface (Plate 34).

- Masonry-type surfaces, such as soft brick, pointing and cement render, may be irreversibly destroyed by grit blasting or excessively high pressure water jetting.
- Plastic glazing and bright metal surfaces, such as stainless steel and aluminium, may be badly scratched by the extensive or repeated use of abrasives, powders or steel wool. The original finish may be difficult to restore and in the case of plastic glazing, good transparency will be lost.

If the wrong graffiti removal agent is employed:

- porous surfaces, such as terrazzo, porous brick or concrete, may be badly stained with graffiti colourants that have been chased into the surface by the mobile solvents that are contained in many graffiti removal agents;
- graffiti marks may be incompletely removed – aerosol paints will not be effectively removed or bleached by bleach-based products (Plate 35).

Three common examples of poor work are as follows.

1. The graffiti removal agent is not properly worked in, particularly on heavily-textured surfaces, or given enough dwell time, so that the graffiti is only partially removed.
2. Unsightly smears remain because the graffiti removal agent is not thoroughly washed off after application.
3. The cleaned area contrasts strongly with the surrounding area where general grime has been left.

Applying conventional obliterative paint over graffiti is undesirable for several reasons.

1. Such areas are often singled out by graffitists for further attacks as a taunt.
2. The obliterative paint is likely to be poorly colour matched with the surrounding paintwork, giving an unsightly patchwork effect (Plate 36).
3. There may be poor compatibility between the original paintwork and the obliterative paint, resulting in bleed-through of graffiti or poor durability.

Obliterative overpainting is never a long-term solution. Frequently-graffitied surfaces should either be cleaned and protected with a good quality surface coating system or overlaid with other material, such as ceramic tile, which has inherent graffiti resistance.

Surface preparation before applying a protective coating is most important. Graffiti, grime and other contaminants, including old paint and the residue of previous treatments, must be removed before the work starts. Failure to do this may result in:

- poor adhesion of the new coating, resulting in limited durability;
- sealing in of residual graffiti and grime which may show through transparent coatings.

15 Health and safety in graffiti removal and surface protection

A major consideration for people who are responsible for graffiti removal and the application of surface protective coatings is the health and safety of those doing the work. Of equal importance is the safety of other members of staff and the public who may be affected by those operations.

In the United Kingdom, everyone concerned with work entailing the use of chemicals is subject to the legal provisions of *The Control of Substances Hazardous to Health (COSHH) Regulations 1988* (see Appendix E for further details). Those regulations apply to any substance which could be hazardous to health, substances listed as having a Maximum Exposure Limit, substances for which the Health and Safety Commission has approved an Occupational Exposure Standard, substantial quantities of dust of any kind and micro-organisms hazardous to health.

The regulations impose duties on employers, and on employees and self-employed persons, for the protection of people who may be exposed to substances hazardous to health at work. They prohibit an employer from carrying on any work which might expose an employee to a hazardous substance unless a full assessment has been made of the risks and of the measures necessary to control exposure to that substance.

The regulations deal in detail with control and monitoring of the exposure of employees to hazardous substances, the maintenance, examination and testing of control measures, health surveillance and the instruction and training of persons who may be exposed to substances hazardous to health.

15.1 Graffiti removal

In graffiti removal, two groups of methods are employed – those that are predominantly mechanical, and those that are essentially chemical.

15.1.1 Mechanical methods

Mechanical methods pose the danger of physical injury. It is essential, therefore, that equipment used for graffiti removal is suitable for the purpose and properly maintained.

Users of the equipment must be fully trained and well instructed in its operation. They must be made to wear protective clothing which may include safety shoes, overalls, gloves, safety glasses, helmets and, where dust concentrations are likely to approach or exceed specified Health and Safety safe levels, respirators.

The nature of dust generated by mechanical graffiti removal methods (e.g. siliceous, metallic, nuisance) should be identified and appropriate respiratory equipment chosen to protect operators against it.

15.1.2 Chemical methods

Three health and safety aspects of chemical methods of graffiti removal should be considered – handling and storage of chemical agents, occupational hygiene and fire hazards.

(a) Handling and storage of chemical agents

A wide range of chemicals are available for graffiti removal purposes. Although many proprietary products have identical formulations, it is important to be aware of the basic ingredients contained in any removal agent used.

Solvent-based graffiti removal agents have de-fatting properties that are liable to cause dermatitis. Skin protection is always needed when such agents are used. Barrier creams should not be used as the primary form of protection. Overalls, gloves and protective visors should be worn. Overalls should be laundered regularly. Solvents of low volatility can accumulate in clothing and when signs of deterioration are seen, overalls should be destroyed.

Protective gloves are commonly made of polyvinylchloride, neoprene, nitrile butyl rubber, polyvinyl alcohol or natural rubber. Each has its advantages and disadvantages in terms of solvent resistance, comfort and cost. Gloves must give protection against the solvent system in use and, at the same time, be flexible enough not to hinder the removal operations. Glove manufacturers should be able to give advice on which gloves are suitable for use with specific solvents.

The greatest risk of solvent splashes arises when graffiti are being removed from overhead surfaces with liquid agents. Because of the dangers of eye irritation or damage, not only should operators be aware of the hazards, properly trained and provided with chemically-resistant eye protection, but a means of irrigating eyes with water should be to hand in case a splash gets through.

(b) Occupational hygiene

There are three routes by which hazardous substances can be absorbed into the body. They may be ingested through the mouth, absorbed through the skin or inhaled. The dangers should be clearly pointed out to everyone engaged in graffiti removal. Smoking, eating or drinking should not be allowed when the work is in progress and the wearing of protective clothing should be strictly enforced.

The most difficult health and safety problem associated with chemical graffiti removal concerns the workplace atmosphere. In internal locations, particularly in confined spaces, fumes from solvents are likely to be inhaled, so cleaning agents of low volatility should be used, provided they can do the jobs for which they are intended.

Where possible, doors and windows should be open when solvent-based removal agents are in use, and in some instances, forced ventilation may have to be provided.

In some cases, suitable solvent blocking or air-fed respiratory equipment may be necessary.

(c) Fire hazards

The wide range of agents used in graffiti removal contains substances of varying degrees of flammability, but all should be treated as presenting a fire hazard. Smoking should be strictly prohibited in graffiti removal areas. Used swabs should be placed in metal bins, removed from the site and safely disposed of. Overalls worn during the removal process should be stored away from possible sources of fire and destroyed if they show signs of deterioration. All substances known to be flammable should be stored in accordance with regulations.

15.2 Surface protection

In surface protection, there are hazards associated with both surface preparation and the application of coatings.

15.2.1 Surface preparation

When mechanical methods of surface preparation are used, such as abrading or grit blasting, the main risk is of physical injury. Equipment, including tools, used for such work must be suitable for the purpose and properly maintained. Operators must be fully instructed in the operation of the equipment and trained in the methods to be employed. They must also be supplied with all necessary protective clothing which may include footwear, overalls, gloves, goggles and respiratory equipment.

Where a great deal of dust is likely to be generated, it may be necessary to minimize it by wetting down, local exhaust ventilation, wearing a dust mask or using manual rather than power tool methods. The nature of the dust should be identified to establish the degree of hazard. Dust will usually come from the abrasive substance (e.g. glass, olivine, grit), the substrate (e.g. brick, concrete, polymers) and any existing paint film (e.g. lead, chrome, organic binders), and other dusts may be present.

In all mechanical surface preparation work, good housekeeping is important. Not only may tidiness prevent physical injury through striking or tripping over equipment, but it may keep dust down.

The precautions necessary when using chemicals to prepare surfaces are no different from those described in relation to chemicals used for graffiti removal. It is prudent to handle all chemicals with gloves and overalls and, if splashing could occur, chemical protective goggles.

15.2.2 Coating application

Protective coatings may be applied by brush, roller or spray. Similar hazards are associated with both brush and roller applications. Spraying is more hazardous than either of the other methods because of the more positive release of solvent vapour, and possibly because the product contains more solvent in the form of thinners.

Nevertheless, many surface-protective substances contain volatile ingredients that may affect operatives even when they are using brushes or rollers. Products containing the lowest possible concentrations of volatile solvents should be used.

The ultimate safety measure is respiratory protective equipment. It should be used only where other controls, such as ventilation or the use of substitute materials, are impracticable, inadequate or have failed. It may be needed with some sprayed two-pack polyurethane systems. These are prone to release free airborne isocyanates. If such systems are used, air monitoring should be carried out and if isocyanates are found to be present, full respiratory protection must be worn. Suppliers of two-pack polyurethane systems should provide detailed information about the composition of the product, and in the UK, Health and Safety guidance notes *EH16* and *MS8* should be consulted (Health and Safety Executive, see Appendix B.1). They give information about toxic hazards, precautions and the medical effects of isocyanates. Operatives regularly using polyurethanes may require periodic medical surveillance.

Residual odours of some surface-protective coatings can be a nuisance in places accessible to the public and the vapours may be hazardous, so it may be advisable to do the work outside the hours when people are most likely to be about, if these solvent-based materials are used.

15.2.3 Fire hazards of coatings

It must be borne in mind that even when the surface-protective coating has dried or cured *in situ*, it may present a serious hazard in the event of a major fire. Under such conditions the rate of heat release can be high and the coating may itself accelerate the spread of the fire by igniting, or it may decompose to produce suffocating or even toxic smoke and fumes.

For this reason, care must be taken when selecting a protective coating for application in any locations where a high fire safety performance will be required, for example, below ground, in high rise apartment stair wells and indeed any place where smoke or fume could quickly accumulate or emergency exits could be restricted.

Realizing this, reputable suppliers of anti-graffiti protective coatings will have their products stringently fire tested against the highest national standards, and they should be able to show independent certification to the user that the product meets the criteria necessary for the special application.

Of course other locations of low fire risk (usually exterior) may not require such stringent fire performance criteria, although national building regulations may well apply.

16 Graffiti prevention – alternative countermeasures

With graffiti, as with all other problems, prevention is better than cure, provided the price is right. In practice, most authorities with graffiti problems adopt a combination of deterrent, protective and removal methods, but because neither deterrent measures nor protective materials and devices will entirely defeat determined graffitists, most of the effort goes into graffiti removal.

Nevertheless, there is a growing tendency to use protective coatings which make graffiti removal easier, and sometimes actually have a deterrent effect. A newly-painted, easily cleaned surface is not an attraction to graffitists. They recognize that their efforts will be wasted because the markings are likely to be removed before many people have noticed them.

16.1 Success in New York

In the New York subway system, once the scene of the most concentrated graffiti epidemic in the world, a new policy of removing graffiti from subway cars and stations without delay and keeping them clean, has resulted in a considerable reduction in graffiti attacks. The numbers of passengers carried has also risen to a new peak, after dropping sharply when nearly all the system's 6200 cars were covered in graffiti.

Part of the policy is never to mix graffitied and clean cars in the same train, and cars in the graffiti-free programme must have any fresh markings removed within 24 hours, or else they are not allowed back in service. New York City Transport's philosophy is that it is better to keep cleaned cars clean and have fewer in service than to let some out still bearing graffiti markings simply to maintain a full service. Cleaned cars are kept in secure locations when they are out of service, and in 1983 the workforce for car cleaning was increased from 825 to 1725.

The New York cleaning programme is expensive, but it is considered to be worthwhile. Increased usage of the system has gone some way to defray the costs.

The cleaning programme is backed up by other initiatives introduced by a special task force known as CAST (Car Appearance and Security Task force) which includes senior managers from various sections of the New York City Transport Authority. Schools have been encouraged to teach children that writing on walls is not clever and that it only hurts others. To support the schools' efforts to discourage children from becoming graffitists, a competition was held for schoolchildren on 'Why I don't like graffiti'. A year-long 'Wipe out graffiti' exhibition was set up in the New York Transit Museum, and a team of 'Graffiti Busters' was formed in a New York City high school to remove graffiti on a voluntary basis.

CAST has also organized increased security at terminals and depots, including police patrols and surveillance by a force of roving Property Protection agents.

16.2 European experience

In a number of cities in Great Britain, the police have urged shopkeepers to restrict the sale of spray paint cans to youngsters. Unfortunately, this has not had much effect as youngsters, refused service at one shop, can always find another where they can get what they want.

Many hundreds of graffitists in London, Paris and other major cities have been arrested and appeared in court. Often, juveniles get away with a caution which most of them regard not as a reproach but as a mark of distinction which enhances their reputation among their peers. Many magistrates regard the perpetration of graffiti as a minor offence. If magistrates realized the extent of the problem and could be persuaded to make offenders pay the costs of removing the graffiti for which they were responsible, and ordered the confiscation of spray cans and other graffiti-producing materials, there might be a reduction in the incidence of graffiti in cities.

In fact, it is possible that some courts are beginning to deal more severely with graffitists. In May 1988, the *Daily Mail* reported that a 'graffiti artist' had been banned from the London Underground by a court. He had been convicted four times in the previous nine months for drawing on posters when he 'became bored' waiting for trains. In March 1989, a 21-year-old man was given a two-year prison sentence and ordered to pay £500 compensation to London Underground for 'causing criminal damage' amounting to £4250 to tube trains and stations.

Video cameras sited in strategic locations can help in the identification of offenders and may have a deterrent effect. Such is the persistence of and determination not to be recognized by some graffitists, however, that the cameras should be installed in the most inaccessible positions and secured with bolts that cannot easily be undone and removed.

Exhortation through 'anti-graffiti' posters has not proved to be an effective means of reducing graffiti. Reward schemes aimed at the public or employees have not been widely tried, but if a way could be found to obtain the cooperation of the public, particularly parents and school teachers, a marked reduction in graffiti attacks would undoubtedly result. The New York initiatives suggest that there is scope for imaginative approaches to educative and competitive schemes that will yield worthwhile results. Indeed, in London, where some school children were invited to help clean a heavily-graffitied underpass, they did so with enthusiasm, and eight months later no more graffiti had appeared there.

Attempts to 'design out' graffiti by covering susceptible surfaces with complex patterns have not proved very successful. Even the most elaborate designs are likely to be marred by tags in colours carefully chosen to contrast with the predominant hues (Plate 37). Nevertheless, more and more organizations are commissioning protective coatings and graffiti-resistant materials in new buildings. Graffiti-resistant paint is being applied to all new train cars, and an increasing number of existing ones, by London Underground, the Barcelona Metro and other rail systems.

Some authorities have found it possible to hide graffiti and frustrate further attacks by planting shrubs and other plants in front of the affected areas. Tough fast-growing bushes and wall-clinging vines or ivy keep the markings out of sight and deny access to would-be graffitists (Plate 38).

In some places where extensive graffiti would conflict with the image the organization wants to project, heavy policing has kept the problem within bounds. Such a place is the Alton Towers Theme Park in the UK, where a substantial security force deters most graffitists, and any graffiti that slips through the net is erased as soon as it is seen. The Alton Towers solution is of course expensive and few authorities are willing or able to solve their graffiti problems by 'throwing money at them'.

In Denmark, where there was a severe graffiti problem, some youth cult figures made an anti-graffiti record and a video which were commercially successful and seemed to be responsible for a reduction in graffiti on trains.

In Barcelona, the incidence of graffiti reached its peak in August 1987. The Metro and the state railway systems were particularly affected. With the Olympic Games scheduled to be held there in 1992, the city authorities embarked on a successful programme aimed at cleaning buildings, clearing rubbish from the streets, eliminating graffiti and presenting Barcelona to the world as a clean, bright, safe, attractive city worthy of the honour of being chosen as a venue for the Games. The problem of graffiti was tackled by:

- a coordinated effort between the transport and city authorities;
- widespread publicity about the problem;
- appeals to local pride;
- the use of persuasion and deterrence as opposed to repression;
- painting all the Metro cars with polyurethane paint;
- imposing on all transport staff the duty to report graffiti as soon as it is seen;
- employing special cleaning crews as well as contractors;
- removing all graffiti and fly-posters within 24 hours;
- where possible, carrying out all cleaning during the day, not only to save money, but to show the public what was being done;
- a continuous search for new cleaning agents and equipment;
- holding meetings and showing films at schools and arranging visits to control centres;
- creating some 'legal' opportunities for graffiti and other damaging activities;
- making parents responsible for payment of fines imposed by courts on offending minors.

These examples of local initiatives suggest that the control and reduction of graffiti can be helped by involving the public, appealing to their civic pride, encouraging community action and making people aware of the considerable costs incurred by their own councils and other authorities in applying graffiti countermeasures. Success in arousing awareness and a change in attitude, even in a very few individuals, could bring about a noticeable and progressive improvement in the appearance of the local environment and help in getting rid of graffiti.

Further reading

These books illustrate and make observations on graffiti, sometimes commenting on their social and psychological causes, and one of them is concerned entirely with cleaning stonework and masonry. They may be of interest for what they show, but only the features by Vuchic and Bata and Wallace and Whitehead deals with graffiti removal or protective measures.

Abel, E.L. and Buckley, B.E. (1977) *The Handwriting on the Wall. Toward a Sociology and Psychology of Graffiti*, Greenwood Press, Westport, Connecticut, USA.

Castleman, C. (1982) *Getting Up – Subway Graffiti in New York*, MIT Press (Publ.), Cambridge, Massachusetts, USA.

Chalfont, H. and Prigoff, J. (1987) *Spraycan Art*, Thames & Hudson, London.

Coopers, M. and Chalfont, H. (1984) *Subway Art*, Thames & Hudson, London.

Cooper Clarke, J. and Charoux, J. (1980) *London Graffiti*, W.H. Allen, London.

Clifton, J.R. (ed.) (1986) *Cleaning Stone and Masonry*, ASTM, 1916 Race Street, Philadelphia, PA 19103, USA.

Huber, J. (1986) *Paris Graffiti*, Thames & Hudson, London.

Poesner, J. (1986) *Louder than Words*, Pandora Press, London.

Rees, N. (1981) *The Graffiti File*, Hazell, Watson & Viney Ltd, Aylesbury, Bucks.

Vuchic, V.R. and Bata, A. (1989) US cities lead fight against graffiti, *Railway Gazette International*, January.

Wallace, J. and Whitehead, C. (1989) *Graffiti Removal and Control*, Special Publication No. 71, CIRIA, Storey's Gate, Westminster, London.

Appendix A
Addresses of manufacturers and suppliers

Important notes on the use of Appendix A

1. The names of manufacturers and suppliers in Appendix A are for information only and in no way does their inclusion constitute a recommendation, approval or committment by London Undergound Ltd, the Author or the Publishers. Conversely, omission of companies from this list does not imply that they will not be able to provide equivalent products and services.
2. The range and composition of products supplied or manufactured by companies herein listed will be changed from time to time and so specific products have not been included.
3. It is the user's responsibility to establish the suitability of any products (chemical agents, protective coatings), and equipment for the particular task in hand and to use them in accordance with manufacturer's instructions and conditions, regulations and laws prevailing at the time.

A.1 Pressure washers

Debequip Ltd, 80 Spencer Road, Belper, Derby DE5 1JW
Tel. 0773 828200

Enviroclean Systems Ltd, South Tyne Works, Haltwhistle,
Northumberland NE49 9DE
Tel. 0498 20566

Neolith Pumps London Ltd, 7 Thames Road, Barking,
Essex IG11 0HN
Tel. 081 591 7799

Warwick Pump and Engineering Co. Ltd, Oxford Road,
Bennsfield, Oxford
Tel. 0865 340322

A.2 Other graffiti removal equipment

General items

The following items are all widely available from wholesale hardware and general cleaning suppliers.

- Swabs (cotton)
- Scrubbing brushes (bristle)
- Scrapers (metal
- Brooms (bristle)
- Buckets (metal)
- Pump applicators (e.g. 'KillaSpray') for water washdown
- Plastic mesh scourers
- Plastic scrapers.

Used swab disposal bins

Suitable metal bins with pedal operated lids are available from:

Wembley Industrial Safety Equipment Co., Unit 25A,
Abbey Manufacturing Estate, Mount Pleasant,
Wembley HA0 1PD
Tel. 081 903 0499

(Appropriate hazard warning signs must be attached to these bins.)

Protective equipment

Protective equipment including gloves, goggles, masks (vapour filter) and other items are available from reputable suppliers who advertise in many Health and Safety trade journals. Advice on suitable types of protective equipment for specific graffiti removal tasks should be sought from them or from independent advisory or regulatory bodies on Health and Safety to ensure that the equipment is suitable for the intended use and conforms to the prevailing national regulations.

A.3 Graffiti removal products

Arrow Chemicals Ltd, Stanhope Road, Swadlincote,
Burton-on-Trent, Staffordshire DE11 9BE
Tel. 0283 221044

Blatchen Ltd, Old House Farm, High Bar Lane, Thakeham,
West Sussex RH20 3EH
Tel. 0798 812844

Chemsearch (UK) Ltd, Landchard House, Victoria Street,
West Bromwich, West Midlands B70 8ER
Tel. 021 525 1666

Chemtec Ltd, Biro House, Stanley Road, South Harrow,
Middlesex HA2 8UW
Tel. 081 864 7379

Croda Paints Ltd, Bankside, Hull HU5 1SQ
Tel. 0482 41441

Crown Industrial Products, Hadrian Works, Holtwhistle,
Northumberland NE49 0HF
Tel. 0498 20421

Crowner Products Ltd, Industrial Hygiene Division, Bar Lane,
Roecliffe, Boroughbridge, N. Yorks YO5 9LS
Tel. 0423 323 226

Dacrylate Paints Ltd, Lime Street, Kirkby-in-Ashfield,
Nottingham NG17 8AL
Tel. 0623 753845

Deb Ltd, 80 Spencer Road, Belper, Derbyshire DE5 1JX
Tel. 0773 822712

DeSoto Titanine PLC, Darlington Road, Shildon,
Co. Durham DL4 2QP
Tel. 0388 772541

Dimex Ltd, 116 High Street, Solihull, West Midlands B91 3SD
Tel. 021 704 3551

Fox Valley Systems Ltd, Dept 3040, Nelson Way,
Nelson Industrial Estate, Cramlington,
Northumberland NE23 9BL
Tel. 0670 713440

Graffiti Management, 6 Dartmouth Road, Ruislip,
Middlesex HA4 0DB
Tel. 0895 639409

Graffittection Ltd, PO Box 13, Robertsbridge,
East Sussex TN32 5PY
Tel. 0424 210915

Hydra Research, 308 Wellingbrough Road, Northampton,
Northants NN1 4EP
Tel. 0604 230779

Janchem Ltd, Centre 21, Manchester Road, Warrington WA1 4AW
Tel. 0925 837744

T. J. & S. Jenkinson Ltd, Susan House, Borron Road,
Industrial Centre, Newton-Le-Willows WA12 0EW
Tel. 09252 21666

Kalon Chemicals Ltd, Bassington Industrial Estate, Cramlington,
Northumberland NE23 8AD
Tel. 0670 713411

3M (UK) Ltd, Elms Industrial Estate, Hudson Road,
Bedford MK41 0HR
Tel. 0234 26 8868

Macpherson & Co. Ltd, Trade Division, Bury,
Lancashire OL10 2RG
Tel. 061 764 6030

Mainsol (UK) Ltd, PO Box 60, Northampton NN4 9JN
Tel. 0604 764267

Modern Maintenance Products Ltd, Bilton Court, Wetherby Road,
Harrogate, Yorkshire HG3 1LN
Tel. 0423 889441

Performance Chemicals, 60 Verney Road, London SE16 3DH
Tel. 071 231 3737

Planlight Chemicals, Lonsdale Road, Dorking, Surrey RH4 1JP
Tel. 0306 887908

Prodorite Ltd, Eagle Works, Wednesbury,
West Midlands WS10 7LT
Tel. 021 556 1821

Regal Chemicals Ltd, 34–38 Steele Road, London NW10 7AS
Tel. 081 965 8191

Remchem Ltd, Unit K, Harlow House, Shelton Road,
Willowbrook Industrial Estate, Corby,
Northants NN17 1XH
Tel. 0536 205562

Servochem, Winterstoke Road, Weston-Super-Mare,
Avon BS23 3YS
Tel. 0934 25421

Stewart Wales Somerville Ltd, Glenburn Road, College Milton,
East Kilbridge, Glasgow G74 5BW
Tel. 03552 22101

Stingel Chemie, UK Distributor, Casdron Enterprises,
12–14 City Road, Winchester, Hampshire SO23 8SD
Tel. 0962 841523

Tensid (UK) PLC, Scandec House, Pyrcroft Road, Chertsey,
Surrey KT16 9HP
Tel. 0932 564 133

Unicorn Chemicals Ltd, Mowbury Drive, Blackpool, Lancashire
Tel. 0253 36101

Valley Industrial Products Ltd, Hastingwood Trading Estate,
Harbet Road, Edmonton, London N18
Tel. 081 884 3402

T. & R. Williamson Ltd, 36 Stonebridgegate, Ripon,
N. Yorks HG4 1TP
Tel. 0765 707711

A.4 Graffiti-resistant coatings

Permanent coatings

Croda Paints Ltd, Bankside, Hull HU5 1SQ
Tel. 0482 41441

Dacrylate Paints Ltd, Lime Street, Kirkby-in-Ashfield,
Nottingham NG17 8AL
Tel. 0623 753845

DeSoto Titanine PLC, Darlington Road, Shildon,
Co. Durham DL4 2QP
Tel. 0388 772541

Fosroc CCD Ltd, Stafford Park, Telford, Salop TF3 3AU
Tel. 0952 290221

Graffittection Ltd, PO Box 13, Robertsbridge,
East Sussex TN32 5PY
Tel. 0424 730291

Prodorite Ltd, Eagle Works, Lea Brook Road, Wednesbury,
West Midlands WS10 7LT
Tel. 021 556 1821

Trimite Ltd, Special Coatings Division, Bush Road, Cuxton,
Rochester, Kent ME2 1HD
Tel. 0634 724422

Tor Coatings Ltd, Portobello Industrial Estate, Birtley,
Co. Durham DH3 2RE
Tel. 091 410 6611

Sacrificial coatings

Tensid UK PLC, Scandec House, Pyrcroft Road, Chertsey,
Surrey KT16 9HP
Tel. 0932 564133

Sealocrete Ltd, Binns Close, Coventry CV4 9WE
Tel. 0203 964567

Appendix B
Advisory and official organizations

Note: Addresses of advisory and official bodies are given for information only. Many of these organizations may not have a dedicated advisory service and it is suggested that courteous written enquiries are likely to receive more sympathetic consideration than telephone calls.

B.1 Health and safety

B.1.1 United Kingdom

- Health and Safety Commission, Baynards House,
  1 Chepstow Place, London, W2 4TF
  Tel. 071 243 6000

- Health and Safety Executive, Room 365A, Baynards House,
  1 Chepstow Place, London, W2 4TF
  Tel. 071 243 6000

- Royal Society for the Prevention of Accidents (RoSPA),
  Cannon House, Priory Queensway, Birmingham
  Tel. 021 233 2461

- British Safety Council, 62–64 Chancellor's Road,
  Hammersmith, London, W6 9RS
  Tel. 081 741 1231

- Institution of Environmental Health Officers, Chadwick House,
  48 Rushworth Street, London, SE1 0QT
  Tel. 071 928 6006

B.1.2 Overseas

- *Austria*
  Federal Ministry of Social Administration,
  Central Labour Inspectorate, Stubenring 1, A1010 Vienna

- *Belgium*
  Nationale Arbeidsraad, Blijde Inkomststraat 17–21, Brussels
  Tel. 02 7354000

- *Canada*
  Labour Canada, Occupational Health and Safety Branch,
  Ottawa, Ontario K1A 0J2
  Tel. 819 997 3520

- *Denmark*
  Directorate of the Danish Labour Inspection Services,
  Rosenvaengtalle 16–18, 2100 Copenhagen Ø
  Tel. 01 382800

Direktoratet for Arbejdstilsynot – Health and Safety Executive,
Landskronagade 33, 2100 Copenhagen Ø
Tel. +45-31-180 088

- *Finland*
  National Board of Labour Protection, Hämeenkatu 13 bA,
  3310 Tampere 10
  Tel. 31 37411

- *France*
  Inspection du Travail, 391 Rue Vaugirard,
  5e Arondissement, Paris
  Tel. 01 8286311

- *Germany*
  Arbeitsambt Düsseldorf, Josef Gockelnstrasse 7,
  4000 Dusseldorf
  Tel. 211 43061

- *Greece*
  Section de la securite du Travail, Rue Pireos 40, Athens

- *Ireland*
  Industrial Inspectorate, Department of Labour, Ansley House,
  Mespil Road, Dublin 4
  Tel. 01 765861

- *Italy*
  Inspectorat médical central du Travail, Via XX de Settembre,
  97/c Rome

- *Japan*
  CIS Centre – Chuo Rodosaigai, Boshi Kyokai (JISHA)/Japan,
  Industrial Safety and Health Association (Affiliated to
  Ministry of Labour), 5-35-1 Shiba, Minato-ku, Tokyo 108
  Tel. 03 452 6847

- *Netherlands*
  General Directorate of Labour, Balen van Andelplein 2,
  Voorburg
  Tel. 070 694001

- *Norway*
  Directorate of Labour Inspection, P.O. Box 8103, Oslo
  Tel. 02 469820

- *Spain*
  Plan Nacional de higiene y seguriad del Trabajo,
  Ministerio del Trabajo, Calle Torrelaguna, Madrid

- *Sweden*
  National Board of Occupational Safety and Health,
  S 17184 Solna
  Tel. 8 7309000

- *Switzerland*
  Arbeitsanstalt Zürich, Pirmdurfstrasse 83, 8000 Zurich

- *USA*
  Department of Labor, 200 Constitution Avenue NW,
  Washington DC 20210, USA
  Tel. 1 202 523 8165

  Food and Drug Administration (FDA), Park Lawn Building,
  5600 Fishers Lane, Rockville, Maryland 20857, USA
  Tel. 1 301 443 3170

  Occupational Safety and Health Administration (OSHA),
  Directorate of Safety Standards Programs,
  US Department of Labor, Washington DC 20210, USA
  Tel. 1 202 523 8061

B.2 Buildings and the environment

B.2.1 United Kingdom

- Building Regulations Advisory Committee, Department of the Environment, Millbank Tower, 21–24 Millbank, London, SW1P 4QU
  Tel. 071 217 4486

- Building Research Establishment Advisory Service, Bucknalls Lane, Garston, Hertfordshire
  Tel. 0923 664664

- Civic Trust, 17 Carlton House Terrace, London, SW1Y 5AW
  Tel. 071 930 0914

- Department of the Environment, 2 Marsham Street, London, SW1P 3EB
  Tel. 071 212 3434

- English Heritage, Fortress House, 23 Savile Row, London, W1X 2HE
  Tel. 071 973 3068

- National Council of Building Material Producers, 10 Great George Street, London, SW1P 3AE
  Tel. 071 222 5315

- Royal Institute of British Architects, 66 Portland Place, London, W1N 4AD
  Tel. 071 580 5533

- Royal Town Planning Institute, 26 Portland Place, London, W1N 4BE
  Tel. 071 636 9107

- Tidy Britain Group, The Pier, Wigan, WN3 4EX
  Tel. 0942 824620

- Town and Country Planning Association, 17 Carlton House Terrace, London, SW1Y 5AS
  Tel. 071 930 8903

B.2.2 Overseas

- *USA*
  US Department of the Interior
  National Park Services, Washington DC

  New York State Department of Environmental Conservation
  Albany, New York

B.3 Transport

B.3.1 United Kingdom

- *Belfast*
  Citybus Ltd, Milewater Road, Belfast, BT3 9BG
  Tel. 0232 351201

- *Glasgow*
  Strathclyde Passenger Transport Executive, Consort House,
  12 West George Street, Glasgow, G2 1HN
  Tel. 041 332 6811

- Liverpool
  Merseyside Transport Ltd (Merseybus), Edge Lane,
  Liverpool L7 9LL
  Tel. 051 254 1254

- *London*
  British Railways Board, PO Box 100, Euston House,
  24 Eversholt Street, London, NW1 1DZ
  Tel. 071 928 5151

  British Waterways Board, Melbury House, Melbury Terrace,
  London, NW1 6JX
  Tel. 071 262 6711

  Docklands Light Railway, PO Box 154, Prestons Road,
  London, E14 9QA
  Tel. 071 538 0311

  Inland Waterways Amenity Advisory Council,
  1 Queen Anne's Gate, London, SW1H 9BT
  Tel. 071 222 4939

  London Buses Ltd, 172 Buckingham Palace Road,
  London, SW1W 9TN
  Tel. 071 222 5600

  London Regional Passengers' Committee, Second Floor,
  Golden Cross House, Duncannon Street, London, WC2N 4JF
  Tel. 071 839 1898

London Underground Ltd, 55 Broadway,
London, SW1H 0BD
Tel. 071 222 5600

Transport 2000, Third floor, Walkden House, 10 Melton Street,
London, NW1 2EJ
Tel. 071 388 0386

- *Newcastle-upon-Tyne*
Tyne & Wear Passenger Transport Executive, Cuthbert House,
All Saints, Newcastle-upon-Tyne, NE1 2DA
Tel. 091 261 0431

B.3.2 Overseas

- *Adelaide*
State Transport Authority, GPO Box 2351, Adelaide,
South Australia 5001, Australia
Tel. +61 8 218 2200

- *Amsterdam*
Gemeentevervoerbedrijf Amsterdam (GVBA),
Prins Hendrikkade 108-144, Postbox 2131,
1000 CC Amsterdam, Netherlands
Tel. +31 20 551 4911

- Barcelona
Transports Muncipals de Barcelona, Calle 60 No 423 Sector A,
Zona Franca, Barcelona 08040, Spain
Tel. +34 3 335 0812/2770

- *Berlin*
Berliner Verkehrs-Betriebe (BVG) – 'West' Berlin,
Postdamer Strasse 188, W 1000 Berlin 30, Germany
Tel. +49 30 2561

Berliner Verkehrsbetriebe (BVB) – 'East' Berlin,
Rosa Luxemburgstrasse 2, Postfach 229, 0 1026 Berlin,
Germany
Tel. +49 2 2460

- *Boston*
  Massachusetts Bay Transportation Authority (MBTA),
  10 Park Plaza, Boston, MA 02116, USA
  Tel. +1 617 722 5000

- *Brussels*
  Société des Transports Intercommunaux de Bruxelles,
  Avenue de la Toison d'Or 15, 1060 Brussels, Belgium
  Tel. +32 2 515 3064

- *Buenos Aires*
  Subterraneos de Buenos Aires (SUBTE), Bartolome Mitre 3342,
  Buenos Aires 1201 DF, Argentina
  Tel. +54 1 89 0631-38

- *Chicago*
  Regional Transportation Authority (RTA), 1 n Deaborn Street,
  Suite 1100, Chicago, IL 60602, USA
  Tel. +1 312 917 0700

- *Copenhagen*
  Hovedstadsområdets Trafikselskab (HT),
  Toftegårds Plads, Gammel Køge Landevej 3,
  2500 Valby, Denmark
  Tel. +45 36 44 36 36

- *Hamburg*
  Hamburger Verkehrsverbund (HVV), Altstädter Strasse 6,
  2000 Hamburg 1, Germany
  Tel. +49 40 30230

- *Hong Kong*
  Commissioner for Transport, Transport Department,
  28/F Queensway Government Offices, 66 Queensway,
  Hong Kong
  Tel. +852 862 3194

  Mass Transit Railway Corporation, PO Box 9916,
  General Post Office, Hong Kong
  Tel. +852 751 2111

- *Lisbon*
  Companhia Carris de Ferro de Lisboa SARL (Carris),
  Rua Primeiro de Maio No 101–103, 1399 Lisboa, Portugal
  Tel. +351 1 363 0266

  Metropolitano de Lisboa (ML), 28 Avenida Fontes
  Pereira de Melo, 1098 Lisboa, Portugal
  Tel. +351 1 575457

- *Lyon*
  Société Lyonnaise de Transports en Commun (TCL),
  PO Box 3167, 19 Boulevard Vivier Merle,
  69212 Lyon Cedex 3, France
  Tel. +33 78 60 25 53

- *Los Angeles*
  Southern California Rapid Transit District,
  425 South Main Street, Los Angeles, CA 90013, USA
  Tel. +1 213 972 6000

- *Madrid*
  Empresa Muncipal de Transportes de Madrid SA (EMT),
  Alcantara 24–26, 28006 Madrid, Spain
  Tel. +34 1 401 3100

  Cia Metropolitano de Madrid, Cavanilles 58,
  28007 Madrid, Spain
  Tel. +34 1 552 4900

  Spanish National Railways (RENFE),
  Avenida de la Guidad de Barcelona 6, Madrid 28007, Spain
  Tel. +34 1 733 6200

- *Marseille*
  Régie Autonome des Transports de la Ville de Marseille (RTM),
  BP 334, 13271 Marseille Cedex 8, France
  Tel. +33 91 95 55 55

- *Melbourne*
  Division of the Public Transport Corporation (The Met),
  60 Market Street, Melbourne 3000, Australia
  Tel. +61 3 610 8888

- *Milan*
  Azienda Transporti Municipali (ATM), Foro Buonaparte 61,
  Milano 20121, Italy
  Tel. +39 2 805 5841

- *Montreal*
  Société de Transport de la Communauté Urbaine
  de Montréal (STCU), 159 rue Saint-Antoine Ouest, Montreal,
  Quebec H2Z 1H3, Canada
  Tel. +1 514 280 5500

- *Moscow*
  Greater Moscow Urban Transit Authority,
  Raushskaja Naberejnaya 22, Moscow 113 127, USSR

  Moskovski Metropoliten, Imeni VI Lenina,
  41 Prospekt Mira, Moscow 12900
  Tel. +7 095 222 1001

- *München*
  Münchener Verkehrs-und Tarifverbund GmbH (MVV),
  Thierschstrasse 2, 8000 München 22, Germany
  Tel. +49 89 2380 30

- *New York*
  Metropolitan Transportation Authority (MTA),
  347 Madison Avenue, New York, NY 10017, USA
  Tel. +1 212 878 7000

  New York City Transit Authority (NYCTA),
  370 Jay Street, Brooklyn, NY 11201, USA

  The Port Authority of New York and New Jersey,
  1 World Trade Center, Suite 67, New York, NY 10048, USA
  Tel. +1 212 466 7000

- *Paris*
  Syndicat des Transports Parisiens (STP), 9 Avenue de Villars,
  75007 Paris, France
  Tel. +33 1 45 50 34 09

Régie Autonome des Transports Parisiens (RATP),
53ter Quai des Grands-Augustins, 75271 Paris Cedex 06, France
Tel. +33 1 40 46 41 41

- *Perth*
  Metropolitan Passenger Transport (Transperth),
  PO Box 6122, Hay Street East, Perth, WA 6000, Australia
  Tel. +61 9 425 2525

- *Philadelphia*
  Southeastern Pennsylvania Transportation Authority (SEPTA),
  714 Market Street, Philadelphia, PA 19106, USA
  Tel. +1 215 574 7300

- *Rome*
  Azienda Tramvie e Autobus del Comune di Roma (ATRC),
  Via Volturno 65, 00185 Rome, Italy
  Tel. +39 6 46951

- *San Francisco*
  San Francisco Municipal Railway (Muni),
  949 Presidio Avenue, San Francisco, CA 94115, USA
  Tel. +1 415 923 6212

  Bay Area Rapid Transit District (BART), PO Box 12688,
  800 Madison Street, Oakland, CA 94604-2688, USA
  Tel. +1 415 464 6000

- *Singapore*
  Mass Rapid Transit Corporation, 25K Paterson Road,
  Singapore 0923
  Tel. +65 732 4433

- *Stockholm*
  AB Storstockholms Lokaltrafik (SL), Box 6301,
  Tegnérgatan 2A, 113 81 Stockholm, Sweden
  Tel. +46 8 786 1000

- *Sydney*
  State Transit Authority of New South Wales, PO Box 1327,
  North Sydney, NSW 2059, Australia
  Tel. +61 2 956 4770

- *Tokyo*
Transportation Bureau of Tokyo Metropolitan Government,
10-1, 2-chome Yurakucho, Chiyoda-ku, Tokyo 100, Japan
Tel. +81 3 32 16 14 11

Teito Rapid Transit Authority (TRTA 'Eidan'),
Teito Kosokudo Kotsu Eidan, 19-6, Higashi Ueno 3-chome,
Taito-ku, Tokyo 110, Japan
Tel. +81 3 38 32 21 11

- *Toronto*
Toronto Transit Commission, 1900 Yonge Street,
Toronto M4S 1Z2, Canada
Tel. +1 416 393 4000

- *Washington*
Washington Metropolitan Area Transit Authority (WMATA),
600 Fifth Street NW, Washington DC 2001, USA
Tel. +1 202 962 1234

- *Vancouver*
British Columbia Transit, 1200 West 73rd Avenue,
Vancouver, BC V6P 6M2, Canada
Tel. +1 604 264 5000

B.4 Cleaning and conservation

B.4.1 United Kingdom

- Association of High Pressure Water Jetting Contractors,
  33 Catherine Place, London, SW1E 6DY
  Tel. 071 828 0933

- Department of the Environment, 2 Marsham Street,
  London, SW1P 3EB
  Tel. 071 276 3000

- English Heritage, Fortress House, 23 Savile Row,
  London, W1X 2HE
  Tel. 071 973 3068

- Historic Buildings and Monuments Commission for England,
  Hampton Court Palace, East Molesey, Surrey, KT8 9AU
  Tel. 081 977 7222

- Historic Buildings Council for Scotland, 20 Brandon Street,
  Edinburgh, EG4 5RA
  Tel. 031 244 2966

- Historic Buildings Council for Wales, Brunel House,
  2 Fitzalan Road, Cardiff, CF2 1UY
  Tel. 0222 465511

- Society for the Protection of Ancient Buildings,
  37 Spital Square, London, E1 6DY
  Tel. 071 377 1644

- Tidy Britain Group, The Pier, Wigan, WN3 4EX
  Tel. 0942 824620

B.4.2 Overseas

- *USA*
  Director Historic Preservation, US Public Building Service,
  General Services Administration, Washington DC 20405

Office of Archaeology and Historic Preservations,
Heritage Conservation and Recreation Service,
US Department of the Interior, Washington DC

Appendix C
Metric/Imperial conversion tables

Pressure

34 bars	= 500 psi
51 bars	= 750 psi
68 bars	= 1000 psi
85 bars	= 1250 psi
102 bars	= 1500 psi
119 bars	= 1750 psi
136 bars	= 2000 psi
170 bars	= 2500 psi

Throughput

500 l/hr	= 1.83 gall/min
600 l/hr	= 2.20 gall/min
850 l/hr	= 3.12 gall/min
1000 l/hr	= 3.67 gall/min
1200 l/hr	= 4.40 gall/min
1850 l/hr	= 6.78 gall/min

Vacuum

50 m bar	= 19.0 ins water lift
100 m bar	= 38.5 ins water lift
150 m bar	= 58.0 ins water lift
200 m bar	= 77.0 ins water lift
250 m bar	= 97.0 ins water lift
300 m bar	= 116.7 ins water lift

Airflow

20 l/sec	= 9.45 cfm
40 l/sec	= 18.9 cfm
60 l/sec	= 28.3 cfm
80 l/sec	= 37.8 cfm
100 l/sec	= 47.2 cfm
120 l/sec	= 56.7 cfm
140 l/sec	= 66.1 cfm
160 l/sec	= 75.6 cfm

Temperature

°C	°F
110	230
100	212
90	194
80	176
70	158
60	140
50	122
40	104
30	86
20	68
0	32
−10	14
−20	−5
−30	−23

Length

1 centimetre (cm)	= 10 mm	= 0.3937 in
1 metre (m)	= 100 cm	= 1.0936 yd
1 kilometre (km)	= 1000 m	= 0.6214 mile
1 inch (in)		= 2.54 cm
1 yard(yd)	= 36 in	= 0.9144 m

Surface or area

1 sq cm (cm^2)	= 100 mm^2	= 0.1550 in^2
1 sq metre (m^2)	= 10 000 cm^2	= 1.1960 yd^2
1 sq km (km^2)	= 100 ha	= 0.3861 $mile^2$
1 sq in (in^2)		= 6.4516 cm^2
1 sq yard (yd^2)	= 9 ft^2	= 0.8361 m^2

Volume and capacity

1 cu cm (cm^3)		= 0.0610 in^3
1 cu metre (m^3)	= 1000 dm^3	= 1.3080 yd^3
1 litre (l)	= 1 dm^3	= 0.2200 gal
1 hectolitre (hl)	= 100 l	= 21.997 gal
1 cu inch (in^3)		= 16.387 cm^3
1 cu yard (yd^3)	= 27 ft^3	= 0.7646 m^3
1 pint (pt)	= 20 fl oz	= 0.5683 l
1 gallon (gal)	= 8 pt	= 4.546 l

Weight

1 gram (g)	= 1000 mg	= 0.0353 oz
1 kilogram (kg)	= 1000 g	= 2.2046 lb
1 tonne (t)	= 1000 kg	= 0.9842 ton
1 ounce (oz)	= 437.5 grains	= 28.35 g
1 pound (lb)	= 16 oz	= 0.4536 kg
1 ton	= 20 cwt	= 1.016 t

US measures

1 US dry pint	= 33.60 in^3	= 0.5506 l
1 US liquid pint	= 0.8327 imp pt	= 0.4732 l
1 US gallon	= 0.8327 imp gal	= 3.785 l
1 short cwt	= 100 lb	= 45.359 kg
1 short ton	= 2000 lb	= 907.19 kg

Appendix D
Solvent properties

Note

The flammability of a liquid is frequently assessed by reference to its Flash Point. This may vary slightly depending upon the method of measurement. Solvents are classified by Flash Points as follows:

Extremely flammable   Flash Point $< 0°$ C, boiling point $35°$ C or lower

Highly flammable   Flash Point $< 21°$ C
Flammable   Flash Point from $21°$ C to $55°$ C
Combustible   Flash Point $> 55°$ C

This particular classification is based on that given in the *Classification, Packaging and Labelling of Dangerous Substances Regulations 1984 (UK)*. Different regulations and classifications will apply in other countries.

Solvent	Boiling point °C	Vapour density	Flash point °C	Occupational exposure limit (ppm)*
Alcohols				
methanol	65	1.11	11	200
n-propanol	97	2.07	15	200
propan-2-ol	80	2.07	12	400
n-butanol	117.5	2.55	35	50
buton-2-ol	99.5	2.55	24	100
n-pentanol	138	3.04	33	—
penton-2-ol	116	3.0	34	—
Ketones				
acetone	56.5	2.0	−18	1000
methyl *iso*-butyl ketone	118	3.45	23	50
heptan-2-one	150	3.94	49	50
Glycol ethers				
2-methoxy ethanol	124	2.6	46	5†
2-ethoxy ethanol	134	3.0	49	10†
2-butoxy ethanol	171	4.1	74	25†
1-methoxy-2-propanol	120	3.1	38	100†
Diglycol ethers				
2-(2-butoxyethoxy)ethanol	230.6	5.58	78	—
Esters				
methyl acetate	58	2.55	−10	200
n-butyl acetate	126	4.0	22	150
pentyl acetate	148	4.5	25	100
iso-amyl acetate	142	4.49	25	100
1-methyl butyl acetate	120	4.48	32	(150)
Glycol ether acetates				
2-ethoxyethyl acetate	156.4	4.72	47	10†

Solvent	Boiling point °C	Vapour density	Flash point °C	Occupational exposure limit (ppm)*
Chlorinated hydrocarbons				
dichloromethane	95–7	6.05	—	100†
1,1-dichloroethane	57–63	3.44	17	200
1,2-dichlorobenzene	180–3	5.05	66	50
Aromatic hydrocarbons				
toluene	110.4	3.14	4	100
xylene (mixed isomers)	138.5	3.66	38	100
1,2,4-trimethylbenzene (all isomers)	168.5	4.1	54	25
Heterocyclic				
N-methyl-2-pyrrolidone	202	3.4	96	100

* Reference has been made to *Guidance Note EH 40/90* issued by the Health and Safety Executive, UK.
(150) This is a short term exposure limit.
† This indicates a maximum exposure limit defined in the above *Guidance Note*. Other values are occupational exposure standards which are concentrations where there is no evidence that exposure is likely to be injurious to employees. Reference should be made to the *Guidance Note* for full definitions.

Appendix E
The COSHH regulations

The *COSHH (Control of Substances Hazardous to Health) Regulations 1988* impose certain legal duties on employers and employees in the United Kingdom, and specify restrictions in relation to certain hazardous substances. The main provisions:

1. require employers not to expose their employees to any substances hazardous to health unless a suitable assessment of risk has been made and safe working conditions are defined;
2. require employees, where practicable, to control health risks at source;
3. state that the employer and employees are jointly responsible for a safe working situation and for ensuring that any protective equipment used is kept in good working condition;
4. requires documentation of maintenance and tests carried out on equipment;
5. may require the monitoring of substances hazardous to health and recording of such monitoring;
6. may require health surveillance of employees exposed to hazardous substances;
7. makes employers responsible for informing and training employees who may be exposed to or may have to work with hazardous substances;
8. lists substances prohibited for certain purposes;
9. lays down the required minimum frequency of examination and test of local ventilation plant used in certain processes;
10. specifies certain substances and processes for which monitoring is required.

The full COSHH Regulations may be obtained from Her Majesty's Stationery Office (HMSO), 49 High Holborn, London, WC1. Tel. 071 211 5656.

Index

Page numbers in **bold** refer to Graffiti Removal Methods (GRMs)

Index

Index